Microvascular Mechanics

Jen-Shih Lee   Thomas C. Skalak
Editors

Microvascular Mechanics
Hemodynamics of Systemic and Pulmonary Microcirculation

With 109 Illustrations

Springer-Verlag
New York Berlin Heidelberg
London Paris Tokyo Hong Kong

JEN-SHIH LEE
THOMAS C. SKALAK
Department of Biomedical Engineering
Health Sciences Center
University of Virginia
Charlottesville, VA 22908, USA

Library of Congress Cataloging-in-Publication Data
Microvascular mechanics: hemodynamics of systemic and pulmonary
  microcirculation/Jen-shih Lee and Thomas C. Skalak, editors.
     p. cm.
   ISBN-13:978-1-4612-8198-6
   1. Microcirculation.   2. Hemodynamics.   I. Lee, Jen-shih.
   II. Skalak, Thomas C.
   QP106.6.M55   1988
   612.1'35—dc20   89-19661

Printed on acid-free paper.

© 1989 by Springer-Verlag New York Inc.
Softcover reprint of the hardcover 1st edition 1989

Typeset by Asco Trade Typesetting Ltd., Hong Kong.

9 8 7 6 5 4 3 2 1

ISBN-13:978-1-4612-8198-6   e-ISBN-13:978-1-4612-3674-0
DOI: 10.1007/978-1-4612-3674-0

Preface

... we do not know a truth without knowing its cause.

Aristotle

Perhaps the greatest hope that may be entertained for a scientific work, whether experimental or theoretical, is that it leads to new thoughts and new avenues of investigation on the part of its readers. In microvascular mechanics, the interplay of rheology, anatomy, and cellular and organ function has only just begun to be addressed. To understand the operational behavior of microcirculation, there is a need to integrate studies at the cellular or molecular level with a quantitative, biomechanical description of the circulatory system. The symposium entitled "Frontiers in Cardiopulmonary Mechanics" held in June 1988 at the University of Virginia was intended to provide a fundamental approach to the description of the circulation from the perspective of microvascular mechanics and to examine new methodology that may advance this effort. This book arose out of the work presented at the symposium.

Aristotle expressed well the need to pursue the causes of a phenomenon in order to achieve a truthful understanding of its nature. In this spirit has each of the quantitative sciences progressed, and in this spirit we hope that this book will provide some understanding of the microvascular events and biomechanical mechanisms underlying the behavior of circulation in general, and of pulmonary and skeletal muscle microcirculation in particular. The integrated treatment of pulmonary and systemic microcirculation provided here is intended to encourage the cross-fertilization of these two research fields. The unique microinvasive experimental methods and systems analysis techniques that have been developed for study of the relatively inaccessible pulmonary circulation may be productively adapted for study of systemic microcirculation. Conversely, the rheological concepts of blood flow in single vessels or vessel networks that have been developed using direct observations in certain accessible systemic microvascular beds may aid in the development of analytical methods for pulmonary vascular research.

The introduction of new experimental and analytical methodologies has often led to new fundamental discoveries, and toward this end the book presents several new methodologies that not only complement established methods, but hold promise for answering some of the outstanding questions in microvascular mechanics. B.W. Zweifach and Y.C. Fung have long provided leadership to the worldwide microvascular mechanics community, and their chapters in this book reflect, first, their research accomplishments to date, and perhaps more importantly, their outlook on the most important unexplored paths that must be followed in the future.

In Chapter 1 Dr. Zweifach provides a perspective on our current understanding of the interaction between the anatomical structure, rheological behavior, and transport properties of the microcirculation and the metabolic needs of the tissues. An evaluation of existing deficiencies in our knowledge is provided, and specific suggestions for the direction of future experimentation are given, including the essential need for data relating determinants of microvascular transport to the metabolic state of the tissue. In Chapter 13, Dr. Fung presents a summary of his elegant mathematical theory of pulmonary circulation, which represents today the most advanced integration of microvascular mechanical events, from the single capillary to the whole organ level. The value of understanding the causes of whole organ events is emphasized, and the utility of a mechanical theory in pointing the way to new experiments is illustrated. In the other chapters of the book, new experimental methods and theoretical work aimed at resolving the basic questions summarized in Chapters 1 and 13 are presented.

Part I is focused on the systemic microcirculation. Chapter 2 discusses the microvascular implementation of indicator dilution methods using intravital microscopy, and explains their role in the study of hemodynamics in complex networks where the measurement of transit times may aid to unravel the functional relationship of different portions of the network. In Chapter 3 a new method for measurement of microvascular flows using particulate flow tracers is described. Application to the myocardial circulation is discussed. Chapter 4 addresses the heterogeneous distribution of red blood cells in microvascular networks, and provides a theoretical computational method capable of predicting hematocrit distribution in specific networks. In Chapter 5 the dispersion of flow and blood cells in microvascular networks is discussed, with an emphasis on the integration of single vessel rheological effects and effects of network topology and geometry. Chapter 6 presents an analytical approach to the biomechanical description of whole organ rheology in skeletal muscle, based on experimental measurements of single vessel mechanics, blood rheology, and whole organ microvascular network anatomy.

Part II is focused on the microcirculation of the lung. Chapter 7 summarizes the means of evaluating the pulmonary capillary blood pressure, the major force responsible for edema formation in lungs. Hemodynamic analyses of the transient arterial and venous pressure due to a rapid vascular occlusion or the injection of a low viscosity bolus are discussed in Chapters 8 and 9. These methods may lead to better quantification of the vasoactivity of pulmonary microcirculation. A density method to estimate the change in the blood volume of pulmonary capillaries using the Fahraeus effect of capillary blood flow and the density fluctuations in the aortic blood is presented in Chapter 10. With the measurement of the dye dilution curves in the arterioles and venules on the surface of the lung, the recruitment of pulmonary capillaries is assessed in Chapter 11. For the modeling of the pulmonary microcirculation, elasticity measurements of the arterioles and venules of the human lung are presented in Chapter 12.

In both Parts, we hope to illustrate the challenge both for the development of quantitative microvascular measurements of blood flow, pressure, transit time, volume, and vessel elasticity, and for the construction of analytical descriptions of organ blood flow. These descriptions may be useful in increasing our understanding of both normal regulatory and pathological behavior of organ systems.

The occasion for the symposium that forms the basis for this book was to honor Dr. Ernst O. Attinger and to celebrate the 20th Anniversary of the Department of Biomedical Engineering at the University of Virginia. We join the symposium participants and all of Dr. Attinger's colleagues in expressing our esteem for the invaluable leadership he has provided in the development of biomedical engineering and cardiopulmonary mechanics both at the University of Virginia and throughout the scientific community. We gratefully acknowledge financial support of the symposium by the University of Virginia, the School of Engineering and Applied Science, and the School of Medicine. We wish to thank Drs. J. Milton Adams and Alex J. Baertschi for their work on the symposium Program Committee, Mrs. Patricia J. Hanson for her help in administrating the symposium, the graduate students in Biomedical Engineering and Mr. Bob Anderson for organizing the symposium activities, and the editorial staff of Springer-Verlag for their assistance in the preparation of this book.

JEN-SHIH LEE
THOMAS C. SKALAK

Contents

List of Contributors

RONALD C. ALLISON
Departments of Physiology, Medicine and Anesthesiology, University of South Alabama, College of Medicine, Mobile, AL 36688, USA

SCOTT A. BARMAN
Departments of Physiology, Medicine, and Anesthesiology, University of South Alabama, College of Medicine, Mobile, AL 36688, USA

THOMAS A. BRONIKOWSKI
Marquette University, Milwaukee, WI 53233, USA

KATHLEEN L. CARROLL
Section of Cardiology, University of Wisconsin Medical Center, Madison, WI 53792, USA

DORIS COPE
Departments of Physiology, Medicine, and Anesthesiology, University of South Alabama, College of Medicine, Mobile, AL 36688, USA

CHRISTOPHER A. DAWSON
Research Service 151A, Zablocki VA Medical Center, Milwaukee, WI 53295-1000, USA

YUAN-CHENG FUNG
Department of AMES-Bioengineering, University of California, San Diego, La Jolla, CA 92093, USA

PETER GAEHTGENS
Department of Physiology, Freie Universität Berlin, D-1000 Berlin 33, Federal Republic of Germany

JOSEPH F. GROSS
Department of Physiology, University of Arizona, Tucson, AZ 85724, USA

JEN-SHIH LEE
Department of Biomedical Engineering, Health Sciences Center, University of Virginia, Charlottesville, VA 22908, USA

LIAN-PIN LEE
Department of Biomedical Engineering, Health Sciences Center, University of Virginia, Charlottesville, VA 22908, USA

JOHN H. LINEHAN
Marquette University, Milwaukee, WI 53233, USA

HERBERT H. LIPOWSKY
Bioengineering Program, The Pennsylvania State University, University Park, PA 16802, USA

COLIN B. MCKAY
Department of Biomedical Engineering, Rice University, Houston, TX 77251, USA

STEPHEN H. NELLIS
Section of Cardiology, University of Wisconsin Medical Center, Madison, WI 53792, USA

AXEL R. PRIES
Department of Physiology, Freie Universität Berlin, D-1000 Berlin 33, Federal Republic of Germany

DAVID A. RICKABY
Zablocki VA Medical Center, Milwaukee, WI 53295, USA

GEERT W. SCHMID-SCHÖNBEIN
Department of AMES-Bioengineering, University of California, San Diego, La Jolla, CA 92093, USA

TIMOTHY W. SECOMB
Department of Physiology, University of Arizona, Tucson, AZ 85724, USA

JUNJI SEKI
National Cardiovascular Research Center, Osaka, Japan

THOMAS C. SKALAK
Department of Biomedical Engineering, Health Sciences Center, University of Virginia, Charlottesville, VA 22908, USA

DONALD W. SUTTON
Department of AMES-Bioengineering, University of California, San Diego, La Jolla, CA 92093, USA

AUBREY E. TAYLOR
Departments of Physiology, Medicine, and Anesthesiology, University of South Alabama, College of Medicine, Mobile, AL 36688, USA

WILTZ W. WAGNER, JR.
Departments of Anesthesiology and Physiology/Biophysics, Indiana University Medical School, Medical Science 374, Indianapolis, IN 46223, USA

RONG-TSU YEN
Department of Mechanical Engineering, Memphis State University, Memphis, TN 38152, USA

BENJAMIN W. ZWEIFACH
Department of AMES-Bioengineering, University of California, San Diego, La Jolla, CA 92093, USA

Part I Systemic Microvascular Mechanics

1
Future Trends in Microcirculation Research

BENJAMIN W. ZWEIFACH

Introduction

The primary function of the terminal vascular bed—the orderly exchange of materials between the blood and parenchymal tissue—is dependent upon the coupling of volumetric flow with the surface area available for exchange. For most tissues, this process involves exchange not only under basal or steady state conditions, but under greatly increased flow rates in line with changing metabolic needs. In structures such as skeletal muscle, a five-10-fold increase in volumetric flow may be required, whereas in others a fairly uniform level of flow is maintained.

The effectiveness of microcirculatory adjustments has usually been evaluated by measuring the capacity to restore tissue volumetric flow rate in the face of a given perturbation (so-called autoregulation) (Johnson, 1986). Although useful for the leverage they provide, overall flow measurements by themselves do not come to grips with the unique contribution of the microvascular system—that is, the orderly distribution of blood within the successive segments of the terminal vascular bed so as to keep the relationships between pressure, flow, and exchange surface area commensurate with a prescribed level of blood-tissue exchange. These adjustments depend both on cellular mechanisms that are the feedback for monitoring disturbances in local homeostasis and on distinctive anatomical features of the hierarchy of microvessels that make up the terminal vascular bed. The latter involve not only the individual branchings, but, overall design characteristics—the number, length, diameter, and branch pattern of the successive segments of the microvasculature.

In general the operational activities of the terminal vascular bed can be considered in relation to three aspects—the delivery of blood, the distribution within the tissues, and the actual exchange between the blood and tissue compartments (Table 1.1). Each of these functional parameters are dependent upon specific organizational features and upon mechanisms for the integration of their activities.

In the main, two approaches have been used to analyze the microcirculatory

TABLE 1.1.   Functional attributes of terminal vascular bed

Delivery
Distribution of cardiac output
(modulate arterial blood pressure via shifts in peripheral resistance)
Distribution
Local adjustment of pressure
(flow rate, transit time)
Exchange
Flow levels versus metabolic needs of tissue
Volumetric flow versus exchange surface area
Intercapillary spacing

system—the conventional physiological methods utilizing measurements of the collective behavior of whole organ systems or appendages (Folkow et al., 1970), as opposed to measurements of the properties of the microvascular network through direct intravital microscopy (Zweifach and Lipowsky, 1984). Measurements of tissue autoregulation have been interpreted in a phenomenological context on the basis of two seemingly separate mechanisms—myogenic (Grände et al., 1977) and metabolic (Berne, 1964). The sine qua non of the microcirculatory system, blood-tissue exchange, has been measured in the main (Renkin and Curry, 1978) by monitoring arteriovenous differences of blood-borne materials across a largely undefined population of capillaries. The 'difficulty of taking into account ancillary changes in the number of capillaries that may be involved leaves unanswered not only exchange surface area relationships but factors such as the associated change in overall resistance, the opening or closure of pathways with substantial differences in conductance, and the distribution of blood cells.

Direct intravital microscopy, on the other hand, provides unique information with respect to discrete segments of the microvasculature, but is of only limited value for interpreting the effectiveness of tissue perfusion and exchange. The availability of extraordinarily sensitive molecular probes that can be used to characterize the family of cells associated with the terminal vascular milieu—endothelial cells (Shepro and D'Amore, 1984), smooth muscle (Somlyo et al., 1969), pericytes (Tilton et al., 1979), and leukocytes (Schmid-Schönbein and Engler, 1987)—has brought to light under in vitro conditions a substantial difference among even specialized cell types, such as endothelial cells or smooth muscle, in terms of their location within the network and in a regional context.

In many ways, the field of microcirculation is at a crossroad. Instead of a cohesive data base on which we can build, we find an ever-expanding array of discrete packets of information from different approaches that cannot readily be assimilated into a well-founded conceptual framework. This difficulty has been compounded in recent years by the rapidly emerging body of information concerned with molecular and cellular events. The direction in which the field moves will be determined by our ability to fit the data into existing concepts of microvascular homeostasis, or the necessity of addressing suitable alternatives. As our information becomes concerned with increasingly smaller subunits down to the molecular level of organization, existing definitions and physical constitutive laws may no longer be applicable.

The complexity and redundancy of these mechanisms makes it necessary at this stage to rely on detailed information obtained from a given tissue or organ, beginning with precise structural and physical characteristics of the network segments as a framework for modeling the system and as a basis for incorporating molecular and segmental data. The broad issues that have been addressed include: (1) organizational features of the network as a whole (Zweifach and Lipowsky, 1984); (2) the mechanism for local modulation and distribution of pressure and flow (Granger et al., 1984); (3) the actual exchange of materials across the blood-tissue barrier (Crone, 1963); and (4) the local feedback by which these relationships are kept in line with changing levels of tissue metabolism (Borgstrom and Gestrelins, 1987).

Delivery and Distribution

There has been a trend to treat the microcirculation as a discrete organic unit. Basic to such an approach is the question of whether a structural framework exists that is essentially the same in all tissues where it serves a nutritional function. Still unresolved is the question of how one can relate the collective activity of large numbers of microcirculatory units in the tissue as a whole to the operational features of single modules or their individual segments.

Much of the broad outlines of our current thinking concerning microcirculatory physiology were already advanced in the classic monograph of August Krogh (1929). Current research is in essence redefining these precepts at a different level of organization (Table 1.2). In recent years, advances in methods for the culture of selected cell types under in vitro conditions and the availability of specific immunologic probes have unearthed a vast reservoir of molecular and cell-related information concerning the basic building blocks of the peripheral circulatory apparatus (Furchgott, 1984). The data have challenged many of our current concepts, but since they are obtained from isolated systems, their relevance to the complex conditions that prevail in the body remains limited.

The classic model of Krogh for transcapillary exchange depicts a 50-μm cylinder of parenchymal tissue around individual capillaries that could be effectively supplied with oxygen by the prevailing blood-tissue concentration gradient. In dealing with the broad range of perfusion levels in skeletal muscle during rest and exercise, Krogh was impressed by the need for a mechanism to recruit and de-recruit capillaries with an active flow during periods of muscle work and proposed some form of active "capillary" control. The issue of capillary recruitment remains unresolved despite a considerable body of negative data (Zweifach and Lipowsky, 1984). The possibility of an active

TABLE 1.2.   Basic problems in current microcirculatory research

Structural microcirculatory module versus functional unit
Intercommunication between endothelial cell, vascular smooth muscle, pericyte, leukocytes
Mechanisms for integrating pressure and flow with exchange
Modulation of physicochemical properties of capillary barrier
In situ rheology of blood cells

contribution of capillary endothelium to modulate lumen vessel dimensions
has received renewed consideration in view of the demonstrated presence of
actin and myosin in most endothelial cells. Observations on liver (Reilly and
McCuskey, 1977) and spleen sinusoids (Ragen et al., 1988) reveal changes in
shape of the endothelial lining cells, both spontaneously and upon stimulation,
that are suggestive of an active role in these specialized organs.

When individual cells along the entire length of capillary vessels in tissues
such as mesentery were stimulated mechanically with a micromanipulator,
they reacted by a shape change but there was no significant effect on the caliber
of these vessels (Chambers and Zweifach, 1944). Detailed examination of the
terminal vascular bed has made it clear that earlier investigators had mis-
takenly extrapolated from observations on narrow terminal arterioles and
precapillaries into a generalized interpretation of capillary vasomotor activity.

Structural Design

An alternative explanation was advanced that took into account the organizational pattern of the terminal vascular network in conjunction with selective
spontaneous opening and narrowing of the arteriolar vessels. A key to the
functional implications of the structural design of the terminal vascular bed
is the deployment of the terminal arteriolar extensions as a continuous,
tapered trunk line from which side-arm twigs are given off repeatedly until
the parent stem is of capillary dimensions. The distal continuation of the
arteriole eventually divides into several capillaries that in turn form the
beginnings of the postcapillary collecting system.

As a consequence of the interconnecting nature of the network of arteriolar
vessels, a number of alternative flow pathways exist from any given location
in the network leading to the capillary meshes. When the entire expanse of
the network is viewed under low magnification, it can be seen that flow is
much more rapid in several of the available flow paths to the capillaries.

The fact that the muscular extensions of the terminal arterioles are positioned as a long stem that is surrounded by the capillary meshes in effect
creates a thoroughfare channel for perfusion that becomes more striking under
conditions where flow is reduced and many of the side branches of the
arteriolar trunk are narrowed. Although the earlier assumption that anatomically distinct channels of this kind are present in all microvascular networks

has been disputed (Hammersen, 1970), the basic principle that should not be overlooked is the arrangement whereby the majority of capillaries originate as abrupt side branches, a feature that allows the distal continuation of the terminal arteriole to carry shunt flow under conditions of both excess and reduced perfusion. This singular feature by itself allows for both spatial and temporal recruitment of capillaries with an active flow.

Early investigators had referred to the junctional configuration formed by these side-arm branchings by the term *precapillary sphincter*, implying a control for perfusion of individual capillary vessels. In many instances, however, changes in the diameter of the more proximal precapillaries can affect flow in as many as 25–40 capillary offshoots.

Recent investigations have placed in perspective the fundamental contribution of arteriolar vasomotion in modulating capillary pressure periodically below prevailing osmotic pressure levels and thereby facilitating blood-tissue fluid balance (Intaglietta, 1983). The feedback for this time-dependent relationship remains a challenge. A number of studies have shown that specific components of the spontaneous vasomotion of terminal arterioles usually originate in the cells at the abrupt, side-arm branching (Colantuoni et al., 1984; Meyer et al., 1988).

Despite the fact that these fundamental processes have been investigated intensively for well over five decades, no consensus has been possible with respect to the operational features on which they depend, largely because of the heterogeneity encountered and substantial regional differences. The multiplicity of related factors that have been brought to light has made it increasingly obvious that a homeostatic balance can be achieved at the microvascular level by any of a number of combinations or permutations.

Although the perfusion of the microcirculatory system is in the main mechanical in nature, there are substantial gaps in our information at the biophysical level that must be dealt with before we can begin to incorporate data concerning cellular mechanisms obtained under arbitrary in vitro conditions into a much more complex in vivo framework (Table 1.2). As a compromise, we are forced to adopt a mechanistic approach so as to be in a position to begin to develop a suitable theoretical approach to this borderline area.

Smooth Muscle Tone

The operation of local homeostatic adjustments implies a feedback loop via the arterioles that is activated when exchange falls below a level compatible with the metabolic status of the tissue. The resulting adjustment of arteriolar smooth muscle is mediated either by a direct modulation of cell tone, or indirectly through other cells such as pericytes or even the endothelial cell. Under normal steady state conditions, smooth muscle cells at each of the hierarchical segments of the network are maintained at a level of basal tone that is characteristic for each tissue (Folkow and Neil, 1971). The level at which this basal state is set represents the ultimate determinant of the responsiveness of the smooth muscle cells to stimuli in general. Still lacking are the details of the mediator pathway that is involved in this type of smooth muscle modulation.

The capacity of endothelial cells to serve as a chemical transducer (Vanhouette et al., 1986) that can modify vasoactive materials during their passage across the endothelium from the bloodstream, or arising within the cell proper, has raised the possibility that endothelium in general may serve as the principal local sensor for disturbances in the tissue milieu. The substantial array of endothelial cells that make up the capillary meshwork could therefore represent the active buffer in the maintenance of arteriolar smooth muscle tone. Criteria to distinguish between these separate sites of microvascular behavior are needed before we can achieve the unique potential of intravital miscroscopy as a tool for determining the relevance of microvascular changes to experimental and disease conditions.

Permeability

The physicochemical features that acount for the unusual permeability of the capillary wall have intrigued investigators for many years. The fact that a membrane of closely apposed living cells could act as a sieve across which the substances were transported in line with their molecular dimensions was against accepted precepts of the permeability characteristics of the plasma membrane of cells and led investigators to conclude that either paracellular pathways were present, or that endothelial cells have a highly specialized porous structure. These postulates have been subjected to detailed analysis on the basis of ultrastructural evidence on the electron microscope level but still remain unsettled.

The existence of aqueous pathways in epithelial and endothelial membranes for the movement of materials between cells had been suggested by earlier workers dealing with tissue injury (Simionescu, 1983). The adhesion of cells in tissue culture to form membranous sheets or tubular arrays with selective permeability was found to be dependent upon the presence of calcium (Chambers and Chambers, 1961). Perfusion of the vascular system in situ with calcium-free solutions led to an increased filtration of fluid that was associated with a weakening of adhesion between contiguous cells and a tendency for the endothelial cells to round up (Chambers and Zweifach, 1940).

We now know that the interendothelial cell spaces contain protein polysaccharide complexes that are a continuum of the materials on the lumenal surface of the capillary endothelium and the outer basement membrane material. In light of currently available information, changes in the perviousness of the intercellular pathways could be interpreted as a modification of the physicochemical properties of the polysaccharide complexes on the outer surface of the endothelial cell (Curry and Michel, 1980).

Although the importance of the endothelial cell in modifying transcapillary permeability has long been recognized, the precise mechanism of this contribution remains largely conjectural. Many musculotropic agents have been found to be associated with an increase in the permeability of the capillary without any obvious change in the dimensions of the vessel. Ultrastructural studies led investigators (Majno et al., 1969) to attribute the increased leakage of macromolecules to a separation of contiguous cell borders due to the contraction of the endothelial cell. Present-day information suggests that the phenomenon is more likely the result of factors that lead to changes in cell shape via cytoskeleton assembly and not to a phenomenon related to a cellular contraction machinery, as in muscle cells.

Macro-Micro Relationships

An especially formidable obstacle for deciphering the complex behavior of the terminal vascular bed has been the heterogeneity that exists at every level of organization. The interpretation and assimilation of the extensive data base that are required to make the information applicable to in situ network conditions have limited our ability to identify the contribution of particular microvascular abnormalities to disease.

A stepping stone in this direction would be a model based on a detailed reconstruction of the microvascular bed in representative tissues for which measurements of selected indices are available at both the macro- and micro-scale level. In conjunction with such a piecemeal analysis, it would be especially informative if the data base included some estimate of the effectiveness of the microcirculatory apparatus in maintaining tissue homeostasis by monitoring levels of typical metabolic by-products other than oxygen per se. An index of microvascular effectiveness would be most useful if it dealt with particular combinations of delivery, distribution, and exchange.

Microcirculation in Humans

To date, intravital microscopy has been of limited value in humans, providing measurements mainly for superficial structures such as the eye (Ditzel and Sagild, 1954) or skin (Fagrell et al., 1984). Investigators have had to rely on indirect methods such as the Doppler shifting of laser light (Mito et al., 1988), or ultrasound (Minamiyama and Yagi, 1988), which provide measurements related to a poorly defined population of large and small vessels. In most studies on humans, the assumption has been made that disease is associated with a parenchymal cell dysfunction that has to be compensated for at the microvascular level by an adjustment of volumetric flow to that tissue. There are many diseases, however, such as diabetes, in which basic aspects of cellular metabolism in general are impaired, so that all cells including those that make up the microvascular apparatus are affected, which thereby interferes directly on an intrinsic level with flow adjustments and exchange.

Information is needed to distinguish between the primary or secondary nature of microvascular modifications under abnormal conditions. Direct measurements of the microvessels in the skin or the eye are extremely useful since they have been found to represent a particularly early manifestation of systemic disease (Davis, 1980). It is hoped that novel methods can be devised (electrical or chemical indices) that will provide information for internal organs of the type obtained in other tissues by intravital microscopy—pressure-flow relationship, resistance, red blood cell flux, vessel dimensions, and transit time.

Acknowledgment.   This work was supported by grant HL-10881 from the U.S. Public Health Service.

References

Berne RM (1964) Metabolic regulation of blood flow. *Circ Res* 14(Suppl 1):1–261.

Borgstrom P, Gestrelins S (1987) Integrated myogenic and metabolic control of vascular tone in skeletal muscle during autoregulation of blood flow. *Microvasc Res* 33:353—376.

Chambers R, Chambers ER (1961) *Explorations into the Nature of the Living Cell.* Harvard University Press, Cambridge, MA, pp 81–87.

Chambers R, Zweifach BW (1940) Capillary and endothelial cement in relation to permeability. *J Cell Comp Physiol* 15:255–272.

Chambers R, Zweifach BW (1944) Topography and function of the mesenteric circulation. *Am J Anat* 75:173–205.

Colantuoni, A. Bertuglia S, Intaglietta M (1984) Quantitation of rhythmic diameter changes in arterial microcirculation. *Am J Physiol* 246:H507–H517.

Crone C (1963) The permeability of capillaries in various organs as determined by the use of the "indicator diffusion" method. *Acta Physiol Scand* 58:292–305.

Curry FE, Michel CC (1980) A fiber matrix model of capillary permeability. *Microvasc Res* 20:96–99.

Davis E (1980) Clinical vasomicroscopy. In Kaley G, Altura BM (eds) *Microcirculation,* Vol III. University Park Press, Baltimore, pp 223–234.

Ditzel J, Sagild U (1954) Morphologic and hemodynamic changes in the smaller blood vessels in diabetes mellitus. II The degenerative and hemodynamic changes in the bulbar conjunctiva of normotensive diabetic patients. *N Engl J Med* 250:587–591.

Fagrell B, Tooke J, Ostergren J (1984) Vital microscopy for evaluating skin microcirculation in humans. *Progr Appl Microcirc* 6:129–140.

Folkow B, Neil E (1971) *Circulation.* Oxford University Press, London, pp 97–124.

Folkow B, Hallback M, Lundgren Y, Weiss L (1970) Structurally based increase of flow resistance in spontaneously hypertensive rats. *Acta Physiol Scand* 79:373–378.

Furchgott RF (1984) The role of endothelium in the responses of vascular smooth muscle to drugs. *Annu Rev Pharmacol Toxicol* 24:175–197.

Grände P-O, Lundvall J, Mellander S (1977) Evidence for a rate-sensitive regulatory mechanism in myogenic microvascular control. *Acta Physiol Scand* 99:432–447.

Granger HJ, Meininger GH, Borders JL, Morff RJ, Goodman AH (1984) Microcirculation of skeletal muscle. In Mortillaro N (ed) *The Physiology and Pharmacology of the Microcirculation*, Vol 2. Academic Press, New York, pp 181–265.

Hammersen F (1970) The terminal vascular bed in skeletal muscle with special regard to the problem of shunts. In Crone C, Lassen NA (eds) *Capillary Permeability*. (Proceedings of the A-Benzon Symposium II.) Academic Press, New York, pp 341–350.

Intaglietta M (1983) Wave-like characteristics of vasomotion. *Progr Appl Microcirc* 3:83–94.

Johnson PC (1986) Autoregulation of blood flow. *Circ Res* 59:483–495.

Krogh A (1929) *The Anatomy and Physiology of Capillaries*. Yale University Press, New Haven, CT.

Majno G, Shea SM, Leventhal M (1969) Endothelial contraction induced by histamine type mediators: An electron microscopic study. *J Cell Biol* 42:647–672.

Meyer JU, Borgstrom P, Lindbom L, Intaglietta M (1988) Vasomotion patterns in skeletal muscle arterioles during changes in arterial pressure. *Microvasc Res* 35:193–203.

Minamiyama M, Yagi SI (1988) Measuring the dimensions of a thin cylindrical vessel by processing ultrasonic reflections with an MEM Cepstrium. In Manabe H, Zweifach BW, Messmer K (eds) *Microcirculation in Circulatory Disorders*. Springer-Verlag, Tokyo, pp 451–456.

Mito K, Ogasawara Y, Hiramatsu O, Wada Y, Tsuijioka K, Kajiya F (1988) Evaluation of blood flow velocity waveforms in intramyocardial artery and vein by Laser Doppler velocimeter with an optical fiber. In Manahe H, Zweifach BW, Messmer K (eds) *Microcirculation in Circulatory Disorders*. Springer-Verlag, Tokyo, pp 525–528.

Ragen DMS, Schmidt EE, MacDonald DC, Groom AC (1988) Spontaneous cyclic contractions of the capillary wall *in vivo*. Impeding red cell flow: A quantitative analysis. Evidence for endothelial contractility. *Microvasc Res* 36:13–30.

Reilly FD, McCuskey RS (1977) Studies of the hemopoietic environment. VI. Regulatory mechanisms in the splenic microvascular system of mice. *Microvasc Res* 13:79–90.

Renkin EM, Curry FE (1978) Transport of water and solutes across capillary endothelium. In Giebisch G, Tosteson DC, Ussing HH (eds) *Membrane Transport in Biology. IV. A, B Transport Organs*. Publisher, City, pp 1–45.

Schmid-Schönbein GW, Engler RL (1987) Granulocytes as active participants in acute myocardial ischemia and infarction. *Am J Cardiovasc Pathol* 1:15–30.

Shepro D, D'Amore PA (1984) Physiology and biochemistry of the vascular wall endothelium. In Renkin EM, Michel CC (eds) *Handbook of Physiology, Sec 2, The Cardiovascular System*. Vol IV, Pt 1. American Physiological Society, Bethesda, MD, pp 103–164.

Simionescu N (1983) Cellular aspects of transcapillary exchange. *Physiol Rev* 63:1536–1579.

Somlyo AV, Vinall P, Somlyo AP (1969) Excitation-contraction coupling and electrical events in two types of vascular smooth muscle. *Microvasc Res* 1:354–373.

Tilton RG, Kilo C, Williamson JR (1979) Pericyte-endothelial relationships in cardiac and skeletal muscle capillaries. *Microvasc Res* 18:325–335.

Vanhouette PM, Rubanyi GM, Miller VM, Houston DS (1986) Modulation of vascular smooth muscle contraction by the endothelium. *Annu Rev Physiol* 48:307–320.

Zweifach BW, Lipowsky HH (1984) Pressure-flow relations in blood and lymph microcirculation. In Renkin EM, Michel CC (eds) *Handbook of Physiology Sec 2, The Cardiovascular System. Vol IV. The Microcirculation* Part 1. American Physiological Society, Bethesda, MD, pp 251–307.

2
Transit Time Distributions of Blood Flow in the Microcirculation

HERBERT H. LIPOWSKY, COLIN B. MCKAY, and JUNJI SEKI

Introduction

The quest for quantitative measures of blood flow in the microcirculation has been a recurrent theme in studies aimed at relating microvascular structure and function in health and disease. To this end, numerous techniques have been developed to measure blood flow in situ within individual microvessels observed by direct microscopy in the living animal. Of the several techniques available, the dual-slit method developed by Wayland and Johnson (1967) has become a valuable tool for measuring the velocity of red blood cells (RBCs) in arterioles and venules ranging in diameter from 100 μm down to the size of the true capillaries, about 5–7 μm. With this technique, and similar methods such as laser-Doppler and optical grating methods, much useful information has been obtained on alterations in flow during physiological events and in light of the viscous properties of blood. A major limitation of such measurements has been their inability to relate microvascular perfusion observed within individual microvessels to the topographical succession of arterioles, capillaries, and venules peculiar to a given tissue. In contrast, the potential to overcome these limitations by using techniques heretofore applied mainly to the determination of regional blood flow appears promising provided that the physical basis of such methods can be fully delineated in the framework of direct observations in the microvasculature. Of particular interest is the classical Stewart-Hamilton technique of indicator dilution, which has evolved into an established method for determination of the mean transit time of blood flow in a well-defined vascular compartment.

Historically, the development of indicator dilution methods represents an interesting example of the evolution of a methodology with advances in technology. An excellent review of these developments is given by Yipintsoi and Wood (1974). In brief, the search for techniques to measure the circulation time of blood in the vasculature focused on the premise that the dilution of an identifiable indicator injected into the circulation could be related to the flow rates of the stream that carried it. Attempts to apply this method first appeared in the 18th century and gained increasing attention during the 19th century with the pioneering experiments of George N. Stewart (1860–1930). By injecting a bolus of hypotonic saline into the vasculature of anesthetized animals, Stewart attempted to quantitate the mean circulation time from measurements of the electrical conductivity of blood sampled in arteries and veins. However, it was not until the refinements introduced by William F. Hamilton (1893–1964) and colleagues that the foundation of the indicator dilution method was laid, which bears the same "Stewart-Hamilton principle" today. As shown by Hamilton's research group, the volumetric flow rate of blood through the circulation (cardiac output) could be directly related to the transient dilution of an indicator injected as a bolus into the arterial stream, and the mean transit time could be deduced from the relationship that vascular volume (V) is equal to the product of volumetric flow rate ($\dot{Q}$) and mean transit time (MTT), where MTT is obtained from:

$$\text{MTT} = \frac{\int_0^\infty t\, c(t)\, dt}{\int_0^\infty c(t)\, dt} \tag{2.1}$$

with $c(t)$ representing the time course of the venous concentration of the indicator. In parallel with developments aimed at refining this method in terms of the choice of indicator, its method and site of injection, and the methods for quantitating its concentration with time, numerous theoretical studies sought to elucidate the mathematical basis for the Stewart-Hamilton technique. The firm mathematical basis of indicator dilution techniques was shown by Meir and Zierler (1954), who analyzed the method in terms of the response of a linear system for which MTT could be related to the volume flows and indicator concentrations with the specific pathways through which the indicator traversed the network. As shown therein, given a system transfer function $h(t)$, which represents the frequency distribution of traversal times for all pathways through a vascular compartment, the average transit time is given by:

$$\langle t \rangle = \int_0^\infty t\, h(t)\, dt \tag{2.2}$$

where

$$\int_0^\infty h(t) = 1.$$

With equivalence of MTT and $\langle t \rangle$, the relationship between vascular volume and the total throughput ($\dot{Q}$) of the vascular compartment can thus be explicitly obtained as:

$$V = \int_0^\infty t\, \dot{Q}\, h(t)\, dt \tag{2.3}$$

Although considerable attention has been given to these techniques in applications to the systemic circulation, either for the determination of cardiac output or regional blood flow in specific organs, less attention has been paid to direct application to studies of the microcirculation. Many studies have employed indicator methods to describe microvascular function by using indirect approaches to the detection of diffusible (from blood to tissue) and nondiffusible indicator dispersion within the microvasculature to delineate the distribution of plasma-borne solutes and their uptake by the parenchymal tissue. A comprehensive review of this literature and the firm mathematical foundation of these techniques is given by Bassingthwaighte (1986). However, with numerous advances in the techniques of intravital microscopy during the last two decades, several attractive applications of the Stewart-Hamilton technique have emerged.

Methods for Determination of Mean Transit Time in the Microcirculation

The basic approach to application of indicator dilution methods to the determination of MTT in the microvasculature has been to introduce a bolus of indicator into a major feeding artery and to observe the transient dilution of the indicator in specific microvessels as it traverses the microvascular network. Apart from the site and nature of the indicator, the major technical problem to be solved is the means by which the concentration of the indicator may be quantitated within individual vessels, for which several choices are at hand. For example, Nellis and Lee (1974) introduced a bolus of cell-free plasma (or saline) as an indicator and employed sensitive photodiodes to measure transient changes in light intensity as the indicator traversed the network. Similar studies by Gaehtgens et al. (1976) employed multiple photodiodes to acquire intensity-time curves (ITCs) as well as measure red cell velocity by the two-slit method. Starr and Frasher (1975) employed finely drawn micropipettes to inject a bolus of Evan's Blue dye into the small arteries bounding a region of tissue (mesentery) and then measured the transients in dye intensity within microvessels by frame-by-frame analysis of cinephotographic films. Baker and colleagues (1979) applied techniques of analog video densitometry to measure ITCs in microvessels for a variety of indicators. By using either fluorescently labeled dextrans (Baker et a., 1979), sulfhemoglobinated RBCs (which appear darker than normal RBCs) (Baker et al., 1980), or fluorescently labeled RBCs viewed under fluorescent microscopy (Baker et al., 1982), the spread of indicator could be viewed throughout successive microvascular divisions following a bolus injection into a small artery. These studies made use of an analog video photoanalyzer (IPM, San Diego, CA) that facilitated measurements of ITCs at two discrete regions of a network as viewed under the microscope.

Common to all of these methods was the ability to obtain ITCs at one or two sites within a network to obtain sufficient information to apply Eq. 2.1 to the determination of MTT from injection to the observation sites, or to calculate the difference in MTT between two specific sites within a tissue. A major advantage of the video techniques employed by Baker et al. (1979) over previous methods using photodiodes was the ability to make video recordings of comparatively large regions of a microvascular network (spanning about 500 μm). Thus, sufficient data could be obtained to calculate MTT between different regions of the network. However the analysis of such measurements required repeated acquisition of ITCs from multiple playbacks of the video recordings. In addition, the analog video techniques employed possessed a minimum spatial resolution of about 4 raster lines of the video image, thus limiting the overall size of the video field that could be explored during the passage of a single bolus of indicator.

With the advent of new techniques for the rapid and real-time digitization of video images and attendant software support for implementation on mini-computers such as IBM compatibles, numerous options are currently available to support the sampling of ITCs at multiple sites over fairly large areas of the microvasculature under low optical magnifications. In a recent study that employed the forerunner of such newly developed methods, techniques of video digitization were employed to acquire ITCs at multiple (six to 10) sites in the microvasculature of cremaster muscle (McKay and Lipowsky, 1988). To illustrate the general approach of this technique, presented in Figure 2.1 is a schematic diagram of the experimental setup employed by McKay and Lipowsky (1988). Video recordings of the traversal of a bolus of fluorescent dye (FITC-dextran) introduced into a contralateral femoral artery were made under epi-illumination fluorescence microscopy of the cremasteric network. Subsequent playback of the recordings was made while digitizing six to 10 "windows" or regions of the video scene at a rate of five to 10 video frames per second. The video digitizer employed was an Eyecom II, Spatial Data Systems, digitizer that provided full-frame digitization into 640×512 pixels, with each pixel possessing 8 bits, or 256 shades of grey. The sensitivity of fluorescence microscopy permitted the use of sufficiently low microscope magnifications such that each pixel could span a region of $1-7$ μm (dependent upon the combination of eyepieces and objectives used) to yield a total video field ranging from 640 μm to 4.5 mm in width as viewed on the video monitor. Although under fluorescence microscopy video sensitivity decreases with decreasing objective magnification (due to diminished numerical aperture), it was found that ITCs could be routinely acquired for regions of tissue spanning 1 mm without any loss in accuracy of the measurement, provided that suitable signal processing techniques were employed.

In view of the limits of video sensitivity for acquisition of ITCs several practical considerations must be made in the choice of analytical methods employed to calculate the MTT from the ITC transient. Based upon the formulation given in Eq. 2.1 it would appear quite reasonable to calculate MTT from measurements of $c(t)$, with satisfaction of the condition that the video scene intensity is linearly related to the dye concentration within a microvessel. This latter constraint has been verified by in vitro calibration studies of dye concentration versus intensity in small-bore glass tubes. However, several additional problems may arise, as can be observed by viewing the representative ITCs shown in Figure 2.2A. As shown there for the ITCs

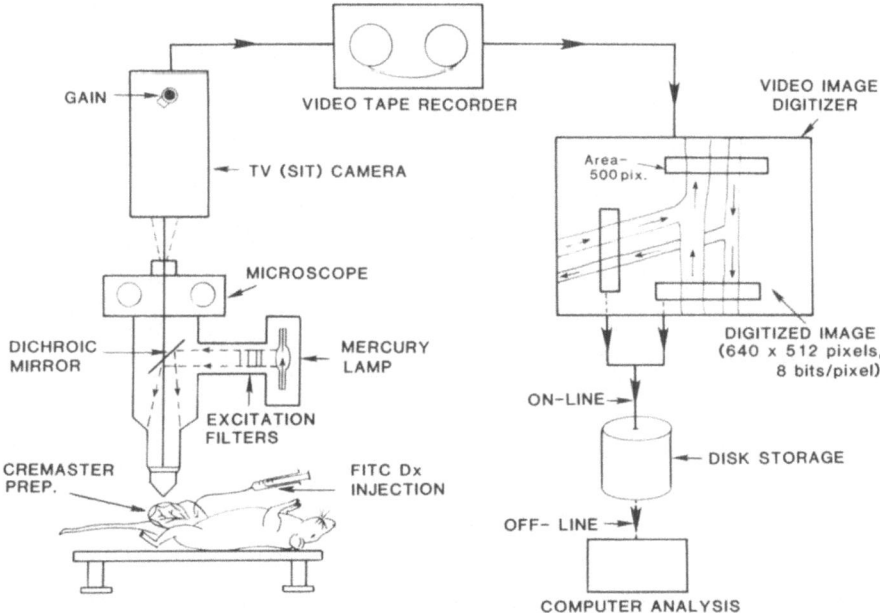

FIGURE 2.1. Schematic drawing of the experimental setup used for the acquisition of intensity-time curves (ITCs) by direct intravital microscopy of the cremaster muscle. Injection of a bolus of fluorescent dye (FITC-labeled dextran; FITC Dx) into the contralateral femoral artery permitted video recordings of ITCs throughout successive microvascular divisions. The ITCs were digitized from videotape recordings and analyzed off line by digital computer. (From McKay and Lipowsky, 1988, with permission of Academic Press, Inc.)

of an arteriole (I_A) and venule (I_V), the precise definition of the limits of integration in Eq. 2.1 and the true zero intensity level of each signal may be obscured by the noise levels inherent to low light level microscopy, or differences in sensitivity of the video camera in different regions of the video scene. To circumvent errors due to these uncertainties, techniques of linear systems analysis were employed to deduce the MTT by cross-correlation of the two ITCs. As suggested by Marmarelis and Marmarelis (1978), MTT was obtained as the time at which a maximum value occurred in the cross-correlation function, $R(\tau)$ of $I_A(t)$ with $I_V(t)$, as given by:

$$R(\tau) = \int_0^\infty I_A(t)\, I_V(t + \tau)\, dt \qquad (2.4)$$

A representative cross-correlation function, $R(\tau)$, is illustrated in Figure 2.2B for the ITCs shown. In general, equivalence of the time of a maxima of $R(\tau)$ and MTT holds only under specific conditions. Given the equivalence of indicator dilution to that of the function of a linear system, as described by Meir and Zierler (1954), the limits of this approximation can be rigorously

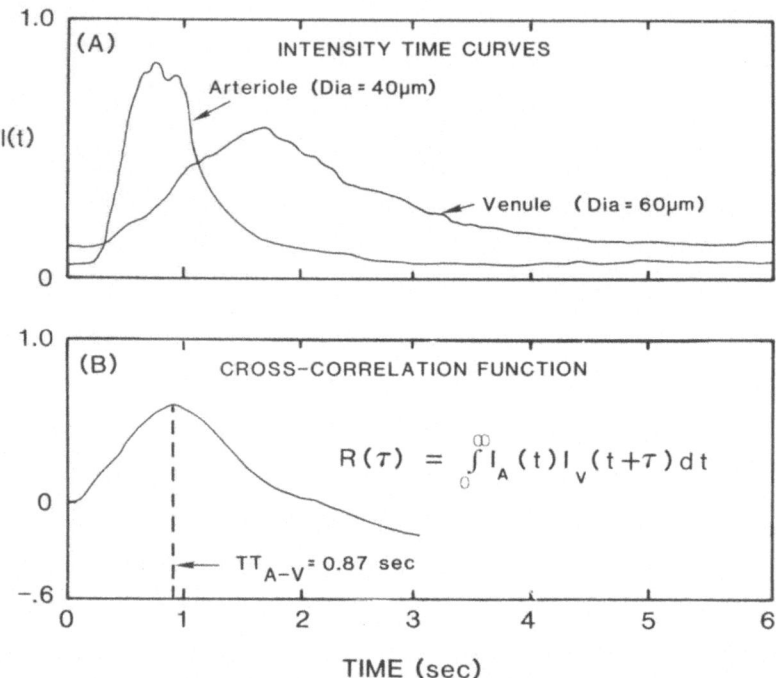

FIGURE 2.2. A: Representative intensity-time curves obtained in an arteriole and venule as a bolus of FITC-dextran traverses the cremaster muscle between these second-order microvessels. The mean transit time between paired arteriole and venule (MTT$_{A-V}$) was determined by cross-correlation of the ITCs. B: The cross-correlation function, $R(\tau)$, corresponding to the curves of panel A is shown. MTT is taken at the time of its peak value, in this case 0.87 s.

detailed. For example, if one treats the input ITCs as the input function of a linear system $x(t)$, and the output ITC as the system output, $y(t)$, then the relationship between the two may be described in terms of the convolution integral:

$$y(t) = \int_0^\infty h(\xi)\, x(t - \xi)\, d\xi \tag{2.5}$$

With application of either Laplace or Fourier transforms, it is apparent from the convolution theorem that the transform $H(\omega)$ may be obtained from $H(\omega) = Y(\omega)/X(\omega)$ and upon inversion of the calculated $H(\omega)$, one may obtain $h(t)$. The average traversal time ($\langle t \rangle$) for all pathways through the vascular network may be obtained from the system transfer function by Eq. 2.2. Suppose that $x(t)$ and $y(t)$ are Gaussian signals, or symmetric about their peaks. It can be readily demonstrated that $R(\tau)$ possesses a maximum at $\langle t \rangle$. To illustrate the applicability of this assumption to transit time determinations made in vivo, presented in Figure 2.3 is a comparison of MTT computed from

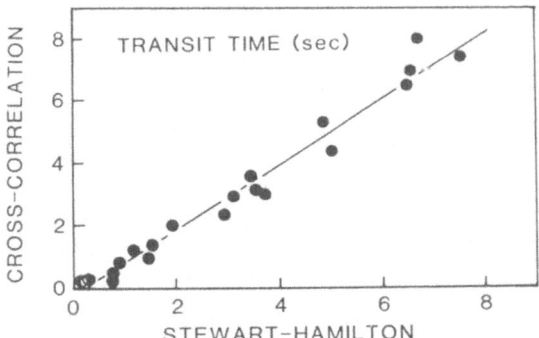

FIGURE 2.3. Comparison of the mean transit time determined by the technique of cross-correlation (ordinate) and that by using the classical Stewart-Hamilton approach (abscissa) (Eq. 2.1). The two methods agreed to within a root mean square error of 13%, thus supporting the use of the simpler cross-correlation method.

the Stewart-Hamilton relationship with that obtained by cross-correlation for 22 measurements of the transit time of a bolus of FITC-dextran from an arteriole to its functionally paired venule in cremaster muscle. It is apparent that the two techniques agree to within a root mean square error of 13%. Thus, the cross-correlation method would appear to offer considerable usefulness in view of its immunity to measurement noise (since broad-band "white" noise in the signals does not affect the computation of $R(\tau)$), simplicity of application, and computational speed (no decisions need be made as to where the d/c components of the ITCs lie).

Applications to the Microvasculature of the Cremaster Muscle

To illustrate the potential for applications of these techniques to relate microvascular structure and function, three examples are considered that attempt to relate MTT to the topography of the microvascular network and the hemodynamics in individual microvessels. First is the utility of indicator dilution methods in defining functionally paired arterioles and venules in light of mass conservation throughout a network, second is the relationship between network topography (vascular order) and MTT, and third is the relationship between network topography and vascular volume.

It appears reasonable to assume that the function of specific segments of the microvasculature may be assessed from the MTT between anatomically identifiable positions in the network. Thus, the MTT between functionally paired arterioles and venules may shed light on the transportive function of a given network division in view of the residence time of blood flow within the division and the time scale for potential transvascular exchange processes.

Although single measurements of MTT from a major artery to specific network divisions may be made and successively subtracted to obtain MTT for a specific division of the network, the averaging of such nonpaired measurements may give misleading information due to spatial heterogeneities. A first step in asserting that indeed functionally paired arterioles and venules can be singled out for observation is to verify that a chosen vascular pair supply and drain the same region of tissue. It is apparent that such decisions made while looking through the microscope are often difficult to substantiate. One useful criterion is the necessary condition that for paired arterioles and venules that service a common region of tissue, the total bulk flow in and out of the intervening region must be identical. In mathematical terms, given a blood-borne indicator with concentration $c(t)$, such as FITC-dextran, the convective flux in an arteriole must equal that in its functionally paired venule (i.e., $\dot{Q}_A c_A(t) = \dot{Q}_V c_V(t)$). Although the measurement of the volumetric flow within each member of an arteriole-venule (A-V) pair is not readily discernible by indicator dilution techniques, the potential for satisfying conservation of the indicator may be validated for each individual A-V pair. To illustrate, shown in Figure 2.4 is the integral of the concentrations (fluorescence intensities) of indicator during the passage of a bolus of dye (FITC-dextran) through a region of tissue bounded by A-V pairs ranging in order from the major arterioles (first order) to the terminal arterioles (fifth order) in cremaster muscle. As indicated by linear regression of the paired measurements, conservation of indicator could be demonstrated with a correlation coefficient of .77 and a slope not significantly different from unity. Hence, although these data lack the necessary verification of mass conservation based upon prevailing bulk flows, they suggest that one can distinguish functionally paired segments of the cremaster network to obtain realistic appraisals of convective transport through a specific anatomical segment.

Additional details of the function of each segment can be obtained from the magnitude of MTT between arterioles and venules of a given segment (order of branching). To illustrate, shown in Figure 2.5A is the distribution of MTT versus order of branching obtained for a representative cremasteric network (McKay and Lipowsky, 1988). The order of branching scheme used here was the centrifugal system where the large arterioles and venules are represented as order 1 and the smallest (terminal or transverse arterioles) are represented by orders 4 and 5. As indicated by the linear regressions of MTT versus order of branching, considerable scatter about the regression is evident. Such scatter is apparently due to spatial heterogeneities throughout the tissue, which diminish considerably within a given tissue as blood flow approaches the level of the true capillaries. As indicated in Figure 2.5B, when MTTs between paired arterioles and venules are plotted as a function of the linear distance from the major entrance of the vascular supply to a tissue, less heterogeneity of MTT is found around a regression line for a given tissue. This situation arises from the fundamental relation that MTT is proportional to the ratio of length scale and mean velocity. We shall return to this point in the *Discussion*.

FIGURE 2.4. The application of indicator dilution techniques to the determination of MTT between functionally paired arterioles and venules of a given order is supported by the apparent ability to demonstrate conservation of the indicator (FITC-dextran). Shown are the time integrals of paired measurements of ITCs in arterioles and venules that were integrated over the duration of the transient. Over the range of five orders of branching in the cremaster network, conservation of indicator is demonstrated reasonably well.

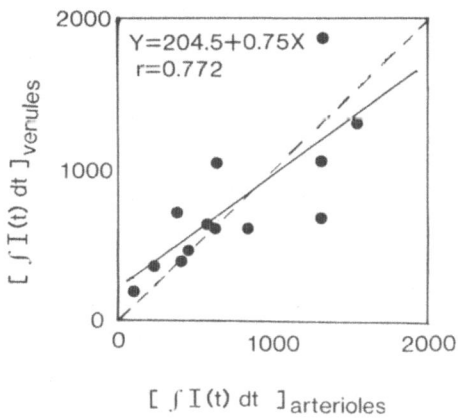

$$Y = 204.5 + 0.75X$$
$$r = 0.772$$

$$[\int I(t) \, dt]_{venules}$$

$$[\int I(t) \, dt]_{arterioles}$$

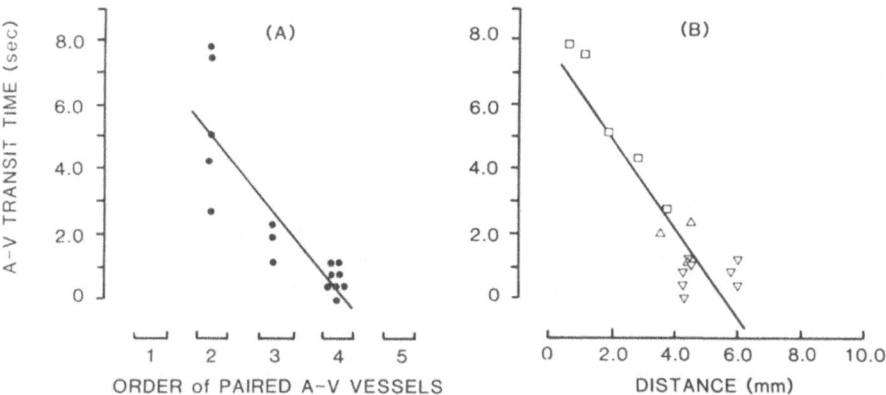

FIGURE 2.5. *A*: The distribution of MTT from paired arteriole to venule is plotted as a function of vessel order for a representative tissue (redrawn from the data of McKay and Lipowsky, 1988). *B*: Regressions with lesser scatter were found when MTT_{A-V} was plotted versus the linear distance along the arterial tree (abscissa) from the hilus of the cremaster to the observation site of the A-V pair. This behavior suggests that MTT may serve as a useful index of the functional position of a constituent microvessel, in contrast to the more subjective assessment of the order of branching.

With knowledge of the MTT of an indicator between functionally paired arterioles and venules it is possible to make use of the basic definition of MTT to determine the microvascular volume (V) served by successive network segments. Given that $V = \dot{Q} \times MTT$, V may be computed from the paired measurements of MTT with the assumption that the total flow ($\dot{Q}$) is the same through all successive divisions. Presented in Figure 2.6A is the average of 76 measurements of MTT determined over the major orders of branching in the cremaster muscle for five different tissues. Using measurements of the bulk

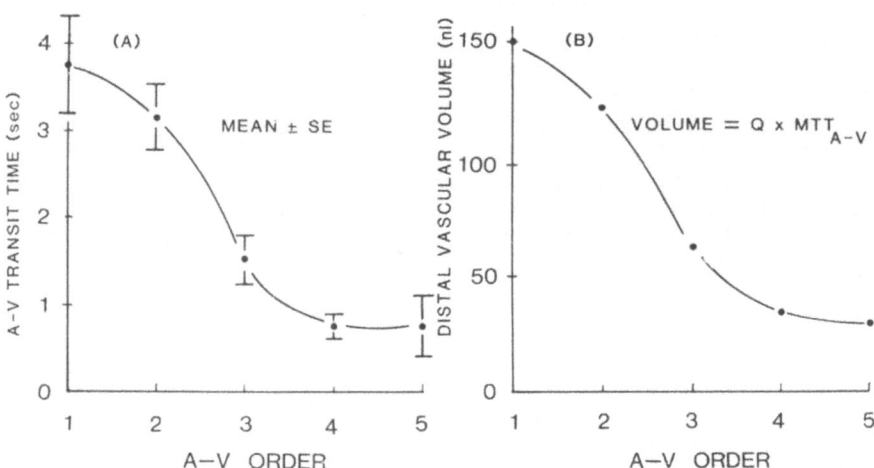

FIGURE 2.6. *A*: The mean transit times (MTT) between functionally paired arterioles and venules are shown for 76 measurements in five tissues, as a function of the order of branching of the cremaster network. With knowledge of the total volumetric flow rate through the tissue (obtained by the two-slit technique) the vascular volume between each A-V pair can be estimated. *B*: The arteriovenous distribution of vascular volume distal to each measurement site, as calculated from the average MTT_{A-V} for each division. Computations such as these may be useful in estimating the rearrangements in network topography found in models of the disease process.

volumetric flow of blood obtained by the two-slit method in the first-order arterioles of cremaster muscle (from House and Lipowsky, 1987), which averaged 40.9 nl/sec, the vascular volume was computed as shown in Figure 2.6B. It should be emphasized that this volume is the total volume distal to the site of observation of an A-V pair at the indicated order of branching. By successively subtracting off the volume of higher orders, one may calculate the vascular volume served per order as 15% for order 1, 45% for order 2, 20% for order 3, and the remainder (20%) for orders 4 and 5 combined. In performing this exercise it was necessary to combine orders 4 and 5, since their close anatomical proximity precluded distinction between transverse arterioles and precapillary vessels, and hence their MTTs are similar.

Discussion

In the present study, we have focused upon the applicability of indicator dilution techniques to direct in situ studies of the microvasculature using the methods of intravital microscopy, with the aim of highlighting the power of these techniques and their potential for future applications. Of the previous applications cited herein, several important features of the relationship between network topography and microvascular function have been obtained. For example, the studies of Nellis and Lee (1974) have demonstrated the substantial departure of the mesenteric microvasculature from a succession of serial elements. Based upon frequency distributions of MTT from superior mesenteric artery to arterioles and venules of various diameters, it was found that considerable overlap of their respective MTTs occurred, thus depicting the extent to which network topography represents an intertwining arrangement of microvessels. While the implications of this topographical feature have yet to be fully appreciated, it is clear that the routes of convective transport of solute to the true capillary level are not strictly ordered throughout a region of tissue.

The studies of Baker et al. (1982) serve as another example of how network topography may vary from one organ to another and the extent to which derangements in topography with pathophysiological conditions may progress. As shown therein, the topography of the cremaster muscle represents a much more ordered deployment of microvessels in terms of an almost dichotomous branching pattern compared to mesentery, as evidenced by the largely monotonic increase in transit time from arterioles to venules. In studies of topographical changes in the spontaneously hypertensive rat cremaster muscle (SHR), Baker et al. (1982) also demonstrated larger values of MTT from feeding artery to small venous vessels. These findings are suggestive of increased vascularity in the venous circuitry of the SHR, and are consistent with prior observations by others who note a rarefaction of arterioles in the SHR model. Reductions in the number of arterioles could give rise to a relatively greater venous transit time compared to that in arterioles, if the arteriolar MTT diminishes due to increased flow rates concomitant with reductions in the number of arterioles.

Based upon the present examples that relate MTT to network topography and vascular volume, several interesting features of these data are apparent that might serve as useful indicators of the state of the microvasculature in various disease models. It is evident from the correlations of the MTT between functionally paired arterioles and venules that the stronger correlation with path length along the vascular tree to the site of the A-V pair, d_{A-V}, suggests that MTT_{A-V} may be a more useful indicator of vascular topography than the order of branching. For example, numerous studies, such as the one by Baker et al. (1982), have attempted to compare hemodynamic parameters between normal and pathophysiological conditions on the basis of order of branching. Given the amount of heterogeneity of MTT within a given order, due to the spatial distribution of apparently similar types of microvessels, the common denominator may not be where such vessels lie in the tissue, but rather the time scale of convective transport to their location in light of the capacity of the vasculature to serve the surrounding tissue by diffusive phenomena.

The fact that MTT_{A-V} correlates more strongly with distance from the major vessel feeding the cremaster muscle to the A-V microvessel pair supports the concept of network design optimization suggested by Murray (1926). As hypothesized therein, the vascular tree may be designed to minimize the energy expended in delivering blood flow to a region of tissue and that the energy dissipated by viscous forces is proportional to the volume of the vascular segment being served. As demonstrated by McKay and Lipowsky (1988), this optimization criterion results in the condition that $MTT_{A-V} \approx k_1 - k_2 \times d_{art}$, where k_1 and k_2 are constants and d_{art} is the linear distance along the arterial tree from feeding vessel to the site of the A-V vessel pair. The applicability of this condition can be demonstrated on the assumption that intravascular wall shear rates vary throughout the vascular tree much less than blood flows vary within individual microvessels of successive network divisions.

In the last example, the distribution of microvascular volume throughout specific anatomical divisions of the cremaster muscle was estimated (Fig. 2.6). It would appear that this parameter is an important variable that may reflect alterations in network topography in the disease state that have been traditionally addressed by morphological techniques. The present approach may offer the advantage of being able to give a realistic functional indicator of the vascular volume within a specific division of a network under specified conditions of vascular tone and regional blood flow. Applications to deciphering the functional significance of vascular derangements in a variety of disease models (e.g., hypertension and diabetes) and blood rheological disturbances may be useful. For example, the present data on volume distribution reveal that about 45% of the vascular volume is dedicated to the second-order vessels, which appear to distribute blood flow spatially throughout the cremasteric muscle tissue, and only 20% of the microvasculature is devoted to the presumably nutritive vessels of the third and fourth orders. Shifts in the relative proportions of these vascular divisions might provide additional insight into the extent to which vascular derangements may participate in specific disease processes. For example, the alterations in vascularity observed in the SHR has at best been the focal point of controversy about the significance of its contribution to the onset of the hypertensive disorder. The acquisition of MTT and vascular volumes throughout specific segments in this case may provide an answer to the question of whether such derangements cause the hypertensive disorder or serve as compensatory microvascular adjustments.

Indicator dilution methods may also represent an ideal method for addressing the rheological behavior of blood flow in the microvasculature. It is generally recognized that the major determinants of the viscous properties of blood in microvessels revolve around blood cell deformability and concentration (hematocrit). In this regard, knowledge of the MTT of a bolus of normal and/or abnormal blood cells throughout successive microvascular segments may provide a unique functional assessment of the determinants of the resistance to blood flow, as well as the extent to which they affect the transportive function of the microvasculature. For example, consider the relationship between microvessel hematocrit (H_m), defined as the packed cell fraction resident within a tube at any instant of time, and the discharge hematocrit (H_D), which represents the volume concentration of red cells actually exiting the tube. It is well known from studies of the Fahraeus effect that H_m/H_D decreases as tube diameter diminishes toward the size of an individual red blood cell. Also, other factors may result in a decreased H_m relative to systemic hematocrit (H_{sys}), such as plasma skimming, stochastic effects, and the departure of the microvessel from its idealization as a cylindrical tube of circular cross-section.

To address the role of hematocrit in microvascular function, one may apply indicator dilution methods to determine the relative volumes within which a bolus of either labeled plasma or red cells traverse the microvasculature. Based upon the definition that H_m equals the ratio of the volume (V) of RBCs to that of RBCs plus plasma, one may obtain:

$$H_m = \frac{V_{RBC}}{V_{plasma} + V_{RBC}} \tag{2.6}$$

Using Eq. 2.3 and the relationships between bulk flow and the volume flow of RBCs ($\dot{Q}_{RBC} = \dot{Q}_{bulk} H_D$) and plasma ($\dot{Q}_{plasma} = \dot{Q}_{bulk}(1 - H_D)$), one may relate H_m to the MTT of plasma and RBCs as:

$$\frac{H_m}{H_D} = \frac{MTT_{RBC}}{MTT_{plasma}(1 - H_D) + MTT_{RBC}H_D} \tag{2.7}$$

From the data of Baker and Wilmoth (1982), measurements of MTT for plasma and RBCs (from superior mesenteric artery to the largest venules in the mesenteric microvasculature of the cat) reveal averages of 7.2 and 5.1 sec, respectively, for which an average value of H_m/H_D may be estimated to be 0.79, given that $H_{sys} = H_D = 0.35$. Similar computations of H_m/H_D for the cremaster muscle of the rat reveal an average value of 0.82.

These calculated regional hematocrits represent a unique view of the effective average hematocrit of each tissue that contrasts markedly from direct observations of H_m within individual microvessels and may thus provide a more realistic appraisal of the extent to which H_m may depart from H_{sys}. In the case of the mesentery (cat), direct measurements of H_m/H_{sys} averaged 0.49 (Lipowsky et al., 1980). For the cremaster muscle, direct measurements of H_m/H_{sys} were found to range from 0.48 in the true capillaries to 0.86 and 0.79 in large arterioles and venules, respectively (House and Lipowsky, 1987). The greater values of H_m/H_{sys} obtained from MTT in these tissues most likely reflect the unique ability of the MTT method to provide an estimate of red cell concentration that accounts for the velocity of red cells through microvessels of a given network division. That is, indicator dilution methods take into consideration the summated effects of the network branching pattern and flows and the attendant apportionment of RBCs and plasma at network branch points. Implicit to the MTT approach is an accounting of mass balance throughout the network, which instantaneous measurements of H_m alone cannot represent. While the potential for such techniques to provide a realistic appraisal of the functional behavior of the microcirculation could stand further reinforcement, they appear promising. Many other areas of study in the transportive function of the microvasculature may form the basis for future studies, as, for example, in the assessment of the relationship between convective transport and transvascular exchange of nutrients and metabolites.

Conclusions

In conclusion, the use of indicator dilution techniques to decipher microvascular function may offer a unique opportunity to quantitate the functional behavior of the microcirculation by the techniques of intravital microscopy. The further refinement of this method, and its application to tissues not readily accessible under transillumination, such as the surface of the brain or other organs, make it an attractive technique. Many potential areas of application of these techniques may be envisaged, which although fraught with numerous technical difficulties, may provide a new approach to understanding the mechanics of microvascular function in health and disease.

Acknowledgment. This work was supported in part by U.S. Public Health Service NIH Research Grants HL-28381 and HL-39286.

References

Baker C, Davis DL, Sutton ET (1979) Arteriolar, capillary and venular FITC-dextran time concentration curves and plasma flow velocities. *Proc Soc Exp Biol Med* 161: 370–377.

Baker C, Davis DL, Sutton ET (1980) Microvessel mean transit time and blood flow velocity of sulfhemoglobin-RBC. *Am J Physiol* 238:H475–H479.

Baker CH, Wilmoth FR (1982) Mesentery and cremaster muscle segmental rbc and plasma volume distribution. *Microcirculation* 2:425–446.

Baker CH, Wilmoth FR, Sutton ET, Takach K (1982) Red blood cell and plasma distribution in SHR cremaster muscle microvessels. *Am J Physiol* 242:H381–H391.

Bassingthwaighte JR (1986) Transport of small molecules across the capillary wall: Assessment via multiple indicator dilution methods. In Baker CH, Nastuk WL (eds) *Microcirculatory Technology.* Academic Press, New York, pp 447–470.

Gaehtgens P, Benner KU, Schickendantz S, Albrecht KH (1976) Method for the simultaneous determination of red cell and plasma flow velocity in vitro and in vivo. *Pflugers Arch* 361:191–195.

Lipowsky HH, Usami S, Chien S (1980) In vivo measurements of "apparent viscosity" and microvessel hematocrit in the mesentery of the cat. *Microvasc Res* 19:297–319.

House SD, Lipowsky HH (1987) Microvascular hematocrit and red cell flux in rat cremaster muscle. *Am J Physiol* 252:H211–H222.

Marmarelis PZ, Marmarelis VZ (1978) *Analysis of Physiological Systems, The White Noise Approach.* Plenum, New York.

McKay CB, Lipowsky HH (1988) Arteriovenous distribution of transit times in cremaster muscle of the rat. *Microvasc Res* 36:75–91.

Meir P, Zeirler KL (1954) On the theory of indicator-dilution methods for the measurement of blood flow and volume. *Am J Physiol* 6:731–744.

Murray CD (1926) The physiological principle of minimum work. I. The vascular system and the cost of blood volume. *Proc Natl Acad Sci USA* 12:207–214.

Nellis SH, Lee JS (1974) Dispersion of indicator measured from microvessels of cat mesentery. *Circ Res* 35:580–591.

Starr MC, Frasher WG (1975) In vivo cellular and plasma velocities in microvessels of the cat mesentery. *Microvasc Res* 10:102–106.

Wayland H, Johnson PC (1967) Erythrocyte velocity measurement in microvessels by a two-slit photometric method. *J Appl Physiol* 22:333–337.

Yipintsoi T, Wood EH (1974) The history of circulatory indicator dilution. In Bloomfield DA (ed) *Dye Curves, The Theory and Practice of Indicator Dilution.* University Park Press, Baltimore, pp 1–19.

3
Particles as Flow Tracers

STEPHEN H. NELLIS and KATHLEEN L. CARROLL

Introduction

Because of the inherent myocardial motion, hemodynamic variables of the coronary microcirculation have been difficult to examine in the intact, freely beating heart. However, some of these variables, such as pressure and diameter, have previously been examined using a free-motion technique (Nellis et al., 1981). With the use of a low-light-level video camera and fluorescent microscopy, we have developed a technique for using particles as flow tracers, specifically to measure the velocity of flow in vessels of the coronary microcirculation.

The free-motion technique used maintains the natural flow conditions of the coronary microcirculation and is a simple procedure that takes advantage of the cyclic nature of heart activity. The heart is visualized for a brief period of time, but at the same time point in each consecutive cardiac cycle; thus the heart appears stationary as viewed through the microscope. In order to do this, it demands that the cardiac cycle, respiration, xenon strobe, and video camera be synchronized.

Methods and Model

The animals used were cats, administered ketamine (30 mg/kg) intramuscularly followed by sodium pentobarbital (30 mg/kg) intravenously via the cephalic vein. The animals were placed in a dorsal supine position on a microscope work table. A tracheal cannula was inserted to allow ventilation by a Harvard model 661 small animal respirator, while the animal was being surgically prepared. A bilateral thoracotomy and sternotomy were done to expose the heart. For the monitoring of pressures, a Millar Mikro-Tip, size 3F, pressure transducer was passed from the right carotid artery into the left ventricle. The jugular vein was cannulated to provide maintenance doses of sodium pentobarbital and the left atrium was cannulated for the direct injection of microspheres. The femoral artery was cannulated for the monitoring of blood pH, P_{CO_2}, and P_{O_2}.

For all synchronization to occur, the heart was paced. The pacer was timed to the video framing rate in such a way that the pace signal occurred at the beginning of a video field with a set number of fields between pacing stimuli.

The particles were fluorescent carboxylated latex microspheres (Polysciences, Inc.). Particles $2\mu m$ in diameter were injected at a constant rate via the left atrium and were illuminated within the vessels by epi-illumination with a Chadwick-Helmuth xenon strobe. The strobe illuminated the heart with three flashes of light in rapid succession. This three-flash burst occurred at the same point in each cardiac cycle; thus the heart appeared stationary. Each particle fluoresced (was seen) at three positions as the particle moved through the microcirculation.

Particles traveling through epicardial vessels have a component of their displacement related to epicardial motion and a component related to the blood flow through the vessel. To subtract the epicardial motion from the total motion of the particle as seen on the image, $8.2\text{-}\mu m$ spheres were injected so that they would become lodged, thereby revealing the motion of the epicardium. In order for a vessel to be measured, at least one of the lodged spheres was located within the field of view.

Figure 3.1 shows the subtraction procedure. The left panel is a video frame showing a particle traveling through an epicardial vessel and a lodged ("stationary") particle. As shown schematically on the right panel, the displacement vectors defined by the lodged sphere are used to correct the vectors of the traveling sphere so that only the component resulting from blood flow through the vessel remains.

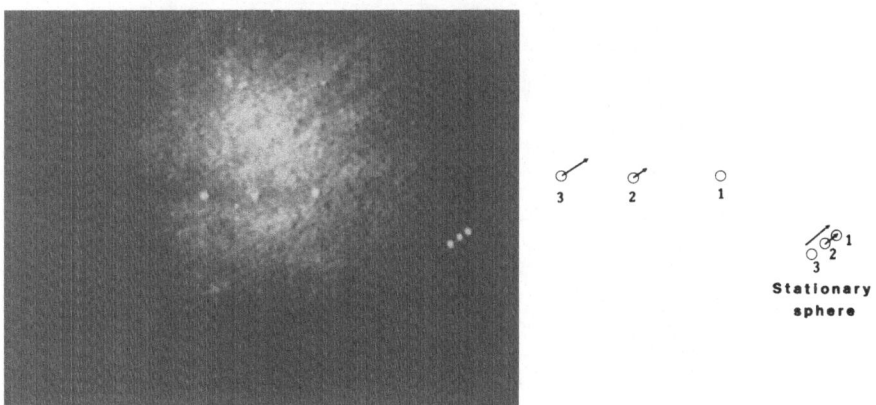

FIGURE 3.1. An example of the vector subtraction of epicardial motion (the lodged "stationary" sphere) from the total motion of a particle as seen on the image of a triple-flash burst.

This approach to measuring flow with fluorescent particles as tracers was somewhat different than previous approaches and therefore we used the rabbit omentum for validation studies of this technique. The illumination of the 2-μm spheres within the vessels was done using the same xenon light. However, because the mesentery was stationary, a continual flashing of the light was used. When looking at the images produced by the continual flashing of the xenon light, there is a "trail" of the particle's positions along its path of motion. An example of an image from the mesentery microcirculation with continual flashing (top) and the coronary microcirculation with the three-flash burst (bottom) is shown in Figure 3.2. The strobe used could operate at frequencies of up to 200 cycles per second.

The time between each flash was known and the distance the particle

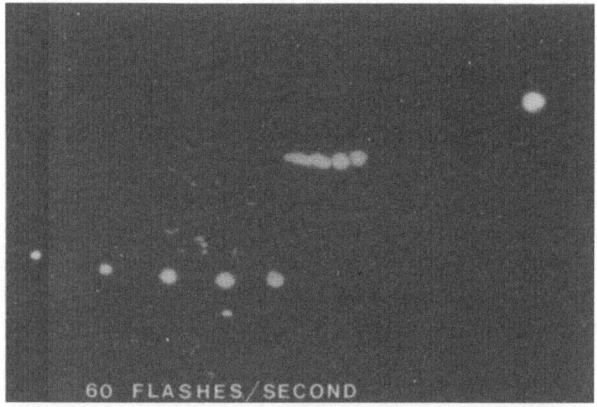

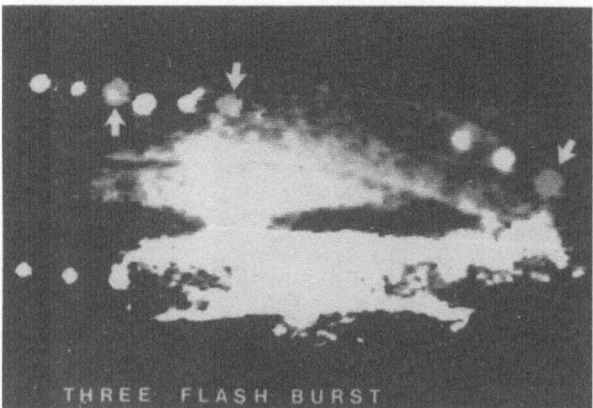

FIGURE 3.2. Photographs were taken of the images produced by the two approaches. The continual flashing of the xenon light is on the top, and the three-flash burst is on the bottom. The *arrows* show the position of the sphere when the third flash of the triple flash occurred.

traveled between each flash could be measured; therefore the velocity of each particle could be calculated. With continual flashing, if the rate of flash was fast enough, the velocity of every sphere that flowed through the vessel segment in a given period of time was measured. With the triple flash, if the vessel segment was positioned correctly and the interval between each of the three flashes was adjusted appropriately, the velocity of every particle that was captured by the flashes was measured.

Velocity Distributions

Although the two techniques were related, there was not a one-to-one correlation between the mesurements. The differences between the two approaches are shown, using a hypothetical case, in Figure 3.3. The continual flashing approach is on the left, while the right depicts the triple flash approach. This hypothetical vessel has only two streamlines. One is traveling at 1 mm/s and the other is traveling at 2 mm/s. Each streamline delivers one sphere per second to the vessel segment, which is 6 mm in length.

If continual flashing is used and the vessel is monitored for 6 s, there would be six spheres traveling at 1 mm/s and six spheres traveling at 2 mm/s; thus the average velocity calculated would be 1.5 mm/s. On the other hand, if the triple-flash image of the 6-mm segment of vessel is used, there would be six of the 1-mm/s spheres but only three of the 2-mm/s spheres seen, so that the calculated average velocity is 1.33 mm/s. The triple-flash approach emphasizes the slower moving particles compared with the continual flashing approach. The velocities measured with the continually flashing xenon light were termed *entrance* or *exit velocities*, and those measured with the three-flash burst were termed *snapshot* or *tube velocities*.

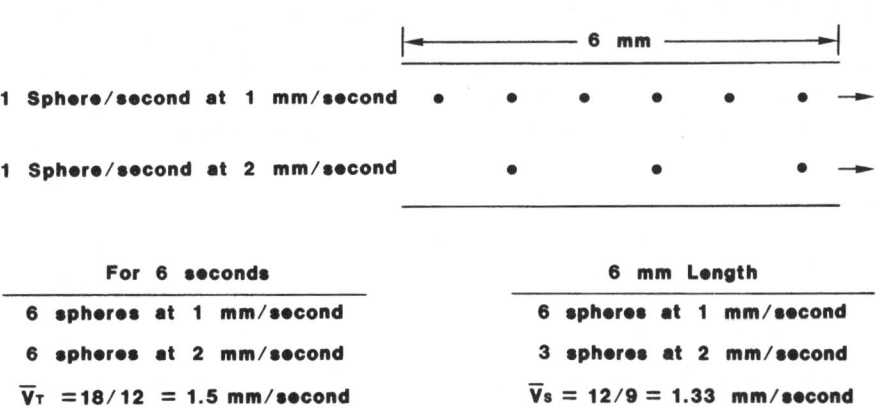

FIGURE 3.3. A hypothetical case illustrating the differences between the continual flash and triple-flash burst. The two approaches result in different calculated mean velocity.

The experiment suggested by the above simple analysis was done and the results are shown in Figure 3.4. The data were obtained from a 100-μm vein in the rabbit omentum. Using continual flashing, the vessel segment was monitored until the velocities of 250 different 2.3-μm spheres were acquired. These data are shown on the left. The data on the right were obtained with the triple flash and again 250 spheres were analyzed. In both cases, the raw data are displayed in the top panel as a cumulative distribution of spheres versus velocity. Both the cumulative spheres and the velocity were normalized to 1. A fifth-order polynomial was fitted to the normalized data. The derivative of this equation (shown in the bottom panel) then served to provide an approximation to the sphere's velocity probability distribution.

The tube velocities show a flat frequency distribution with an equal likelihood of finding a sphere traveling at any of the slower velocities. As expected, the frequency distribution obtained with continual flashing starts low and reaches a peak at the higher velocities. Pulsatile flow and measurement errors distort the distribution curves. This distortion would be most pronounced at the high velocities; therefore, caution should be taken when interpreting this portion of the curves. The average velocities calculated from the two sets of data are marked on the bottom panels.

Analysis of Frequency Distribution

To better understand these frequency distributions it is helpful to examine some idealized examples, such as that of a parabolic flow profile. For this flow, the equation for the normalized velocity of flow in a vessel is

$$V = (1 - r^2/R^2) \tag{3.1}$$

where R is the radius of the vessel. The velocity (V) is 0 at the vessel wall ($r = R$) and increases to 1 at the center ($r = 0$). If a sphere has a velocity larger than V, the sphere will situate at a radius smaller than r when V and r are related by Eq. 3.1. Since the total area of the tube is πR^2, the probability of finding a particle with a velocity greater than V is the ratio of the two areas πr^2 and πR^2. Thus the probability of finding a particle with a velocity between 0 and V is

$$P = 1 - r^2/R^2 \tag{3.2}$$

For the velocity profile given in Eq. 3.1, we have

$$P = V \tag{3.3}$$

subject to the conditions that $P = 0$ with $V < 0$ and $P = 1$ with $V > 1$. The sphere probability distribution $p \, (= dP/dV)$ is thus

$$p = 1(V) - 1(V - 1) \tag{3.4}$$

where $\mathbf{1}(V)$ is the unit step function. We note that $p\, dV$ is the fraction of spheres having a velocity in between V and $V + dV$.

Due to the exclusion of spheres from the regions adjacent to the vessel walls, the velocity distributions for all vessels studied show that the lowest velocity is non-zero. We represent the velocity profile of such flows as

$$V = 1 - (1 - 1/n)r^2/R^2 \qquad (3.5)$$

where n is some constant such that $n > 1$. The minimum velocity in this profile is $1/n$. Then, using Eq. 3.5 for conversion of r^2/R^2 in Eq. 3.2 to V leads to the following cumulative probability of finding a particle with velocity between $1/n$ and 1

$$P = (V - 1/n)/(1 - 1/n) \qquad (3.6)$$

subject to the condition that $P = 0$ if $V < 1/n$ and $P = 1$ if $V > 1$. The corresponding sphere probability distribution is

$$p = [\mathbf{1}(V - 1/n) - \mathbf{1}(V - 1)]/(1 - 1/n) \qquad (3.7)$$

As shown in Figure 3.4, the inclusion of this non-zero lower normalized velocity limit $(1/n)$ yields a better simulation of the measured cumulative probability than that given by Eq. 3.3.

The repeated measurement of spheres traveling at a constant velocity has been carried out using a spinning disk. The standard deviation of measurements so obtained is related to the velocity by this linear equation:

$$s = .01 + .05V \qquad (3.8)$$

The zero velocity intercept is approximately equal to a 1-pixel error in marking the sphere location; therefore, its value as related to normalized velocity depends on the interval between flashes. An error of 1-pixel translates into a larger absolute error in velocity if the flash duration is short. Eq. 3.8 represents the expected error in marking spheres for the flash duration used to acquire the data shown in Figure 3.4.

The distribution of sphere velocities measured is a combination of the actual velocity distribution found in the vessel and the dispersion created by measurement errors. The probability p_m of measuring a sphere as traveling at a velocity V can be calculated by integrating the contribution from all velocities in the vessel. If the dispersion is approximately a Gaussian distribution, the measured frequency distribution is

$$p_m(V) = \int_{1/n}^{1} (2\pi s^2)^{-1/2} \exp[-(V - u)^2/(2\,s^2)]\, du \qquad (3.9)$$

where the relationship between velocity in the vessel u and standard deviation s is given by Eq. 3.8 with V replaced by u. Because of the pixel nature of the images, measured velocities are represented by discrete numbers and not a continuum. This effect is most pronounced at the low velocities. As shown by the solid lines in Figure 3.5, the measurement errors are not sufficient to

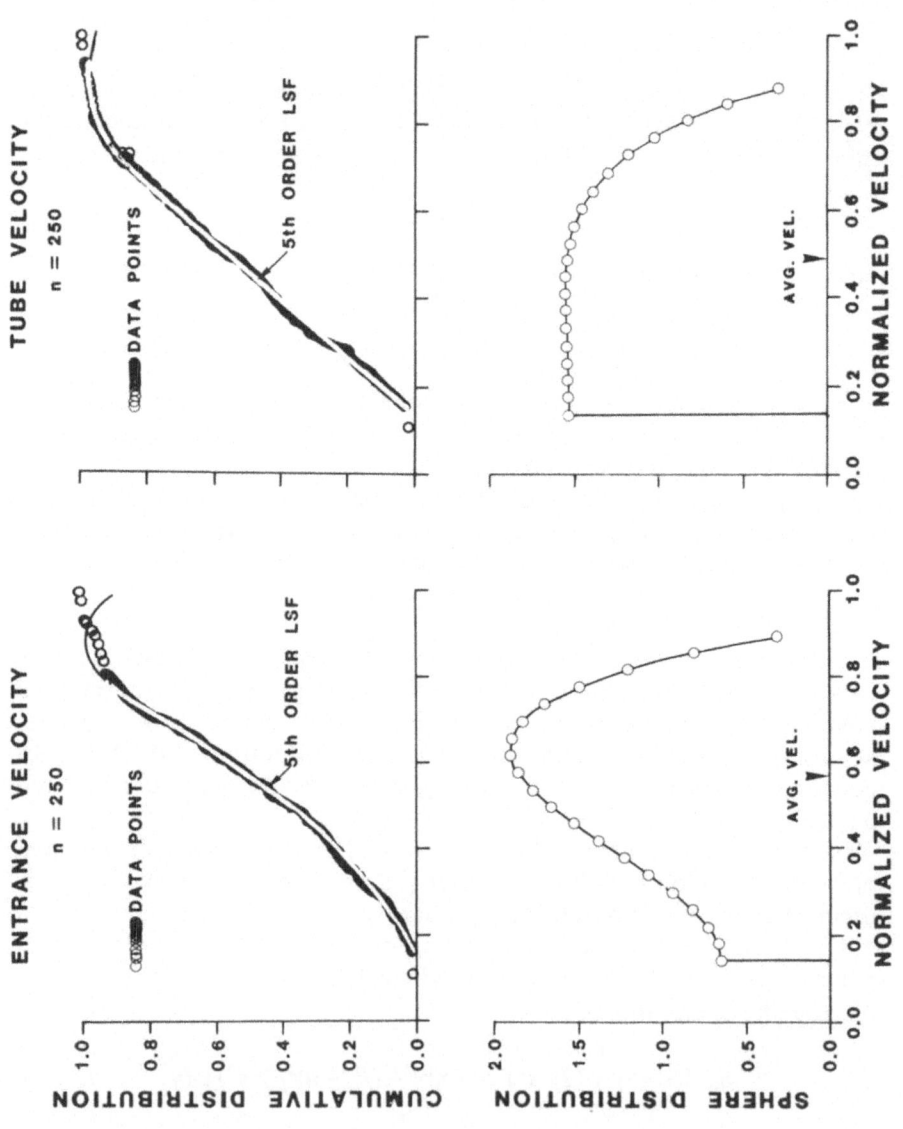

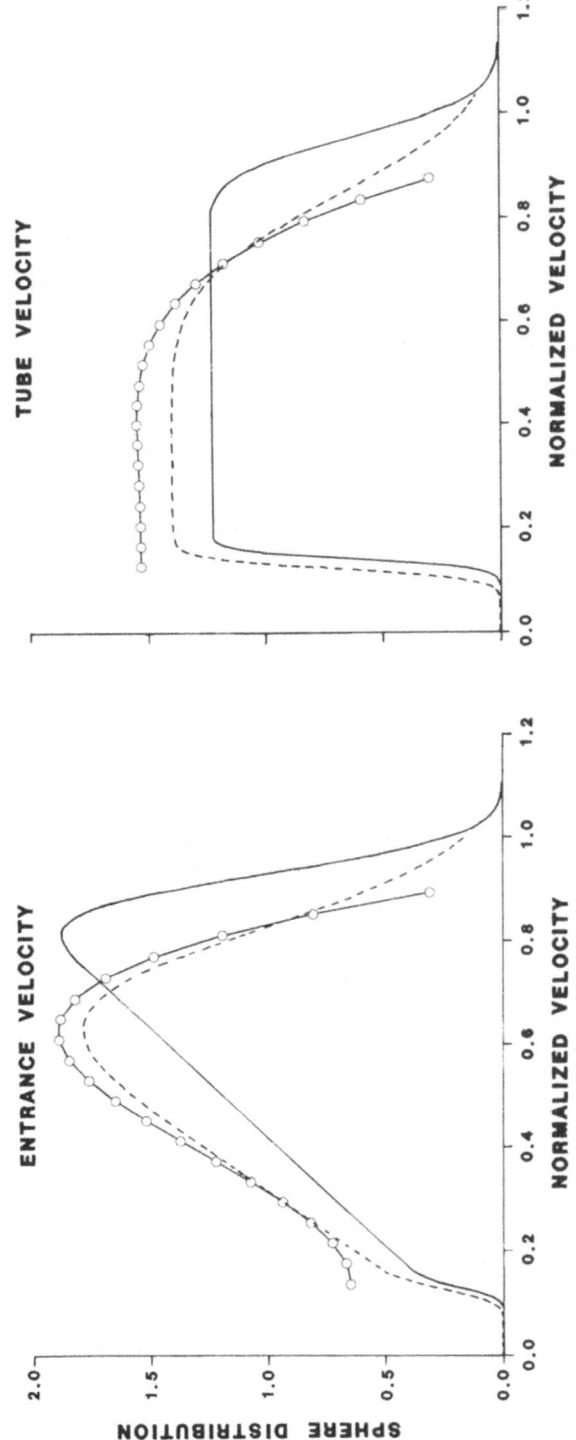

FIGURE 3.4. Data obtained using the continual flash approach and the triple flash approach on a 100-μm vein from a rabbit omentum were compared. Two hundred fifty 2.3-μm spheres were measured. The cumulative distributions of spheres with velocity are shown in the top panels. Velocity is normalized to 1. The frequencies of velocity distributions were calculated from the curve that approximates these data and are shown in the bottom panels.

FIGURE 3.5. The sphere probability distribution shown in Figure 3.4 (○—○ line) is compared with an idealized case of a parabolic velocity profile with measurement errors. The measurement errors for one curve (solid line) were estimated from spheres coated on a spinning disk. The measurement error for the third curve (dashed line) was assumed to be 15% of the mean velocity.

explain the dispersion. Accounting for the error due to cardiac pulsation is a curve (shown as the broken lines in Fig. 3.5) for which the standard deviation is assumed to be 15% of the mean. This percentage was chosen because it results in a curve that better approximates the data from a subjective viewpoint.

The cumulative distribution from the measurement was normalized by the fastest sphere found. In order to compensate for this scaling the distribution predicted by Eq. 3.9 was scaled so that there is a 99% probability that a sphere will be measured at a velocity less than 1. This scaled function is plotted in Figure 3.5. Also included in Figure 3.5 is the probability distribution of Figure 3.4. The panel on the left shows the probability distribution for the case of continual flash.

Concentration and Velocity Profile

The top panels of Figure 3.6 are the cumulative distributions of sphere velocities measured using the triple flash method. The only difference between the two cases shown was the size of the vessels. The data on the left were acquired from a 30-μm vein while the data on the right were from a 100-μm vein. There was a high concentration of spheres traveling at the slower velocities in the smaller vessel. This altered frequency distribution could have resulted if the flow deviated from a parabolic profile or it could have resulted from a heterogeneous radial distribution of the spheres in the vessel.

For 2-μm spheres, an increased number of the spheres travel at slower velocities, probably resulting from a combination of a flat velocity profile and a near-wall excess of the small spheres. Other investigators have reported both phenomena (Tangelder et al., 1985; Tilles and Eckstein, 1987).

Conclusions

This chapter presents a new technique for study of the velocity distribution in microvessels, which can provide information on how microspheres or other particulate flow tracers are transported in the vessels. With this information, the transfer function describing the transport of red cells in microvascular flows can be established and be used for a better assessment of the flow and the average red cell velocity in the microvessel.

Acknowledgment. A National Heart, Lung and Blood Institute grant (HL-39282-02) and the Rennebohm Foundation of Wisconsin supported this work.

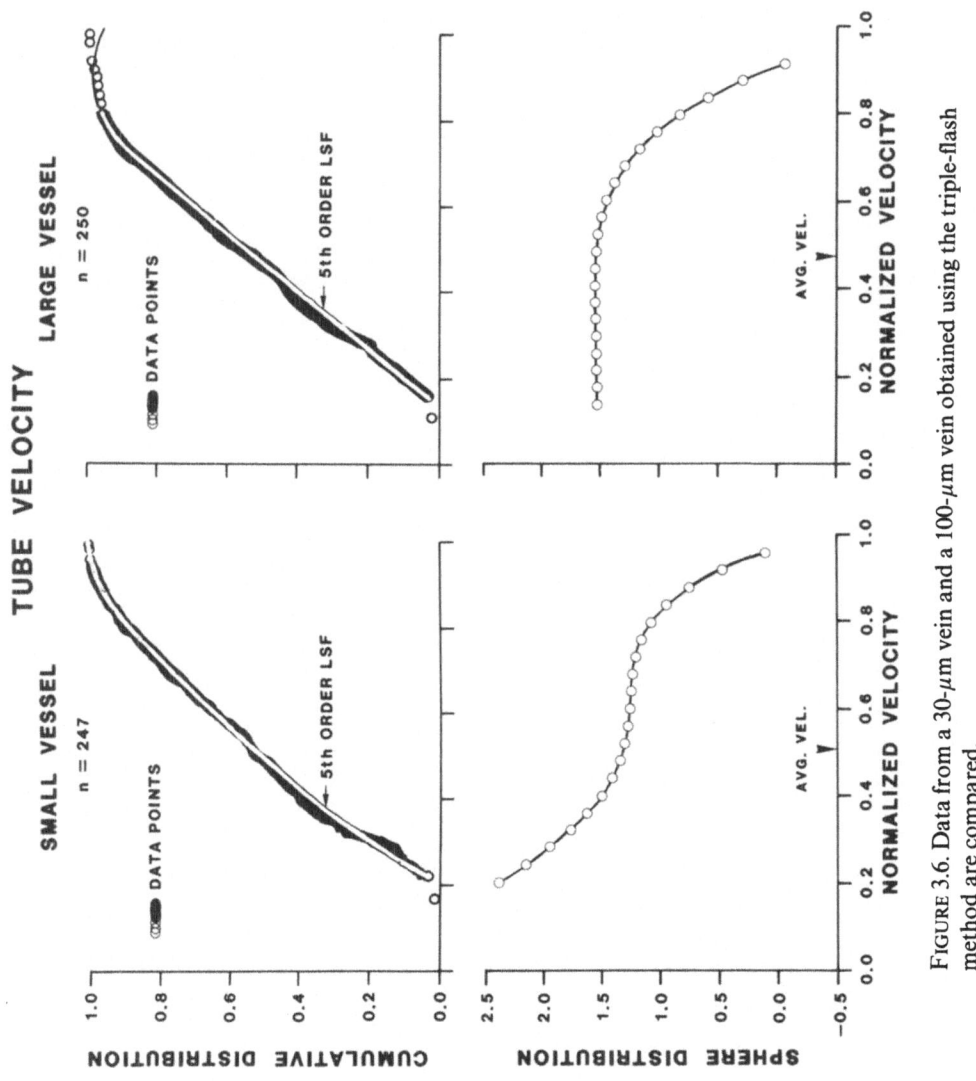

FIGURE 3.6. Data from a 30-μm vein and a 100-μm vein obtained using the triple-flash method are compared.

References

Nellis SH, Liedtke AJ, Whitesell L (1981) Small coronary vessel pressure and diameter in an intact beating rabbit heart using fixed-position and free-motion techniques. *Circ Res* 49:342–353.

Tangelder GJ, Teirlinck HC, Slaaf DW, Reneman RS (1985) Distribution of blood platelets flowing in arterioles. *Am J Physiol* 248:H318–H323.

Tilles AW, Eckstein EC (1987) The near-wall excess of platelet-sized particles in blood flow: Its dependence on hematocrit and wall shear rate. *Microvasc Res* 33:211–223.

4
Theoretical and Experimental Analysis of Hematocrit Distribution in Microcirculatory Networks

Timothy W. Secomb, Axel R. Pries, Peter Gaehtgens, and Joseph F. Gross

Introduction

Many investigators have reported that capillary hematocrits in a variety of tissues are substantially lower than corresponding systemic hematocrits (Johnson, 1971; Klitzman and Duling, 1979; Lipowsky et al., 1980). Part of this apparent discrepancy is due to the Fahraeus effect in individual segments (Fahraeus, 1928), tube hematocrit (volume fraction of red cells within a vessel) being lower than discharge hematocrit (volume fraction of red cells in blood flowing through a vessel). However, *in vitro* measurements and theoretical arguments indicate that this effect alone is not sufficient to explain the observed hematocrit reduction.

Microvascular hematocrits typically vary substantially from vessel to vessel within a tissue. If hematocrits are positively correlated with flow rates in the segments of a microvascular network, it follows from conservation of red cells and plasma that the mean of the vessel discharge hematocrit is lower than the feeding hematocrit. This phenomenon has been termed the *network Fahraeus effect* (Pries et al., 1986). To calculate the hematocrit reduction due to the network Fahraeus effect, the mean discharge hematocrit is computed on a complete flow cross-section, weighted according to vessel cross-section area. A complete flow cross-section is defined as a set of vessels that together carry the complete flow traversing the network. This mean discharge hematocrit, divided by the overall feed hematocrit to the network, yields a ratio that is less than 1 when the network Fahraeus effect is present.

The network Fahraeus effect was estimated in the rat mesentery by Pries et al. (1986) based on photometric hematocrit measurements. The area-weighted mean discharge hematocrit decreased (i.e., network Fahraeus effect increased) from proximal to distal flow cross-sections, including vessels of increasing branching orders. Mean tube hematocrit was found to decrease more rapidly than discharge hematocrit in successive flow cross-sections, and this additional decrease was attributed to the Fahraeus effect within individual segments (the *vessel Fahraeus effect*). Pries et al. (1986) hypothesized that the positive correlation of velocity and hematocrit in the network underlying the observed network Fahraeus effect is accounted for by the tendency of the high flow branch of a diverging microvascular bifurcation to receive a higher hematocrit than the low flow branch (the *phase separation effect*).

The aim of the present study is to determine whether the observed discharge hematocrit distribution in the rat mesentery can be explained in terms of the phase separation effect mentioned above. A theoretical simulation of flow and hematocrit distribution, based on available information on phase separation effect and blood viscosity, is applied to a microvascular network with known geometry in the rat mesentery. Predicted hematocrits in each segment of the network are compared with observed values. The analysis provides predictions of the network Fahraeus effect that are compared with estimates based on measured hematocrits.

Methods

Microscopic Observations of Networks

The theoretical simulations presented here are based on experimental data obtained by intravital microscopy of the rat mesenteric microcirculation. These data include anatomical information (i.e. details of the length, diameter, and interconnections of every segment in the network) together with measured values of hematocrit and information on the flow direction in every segment.

The methods used to obtain the anatomical information have been described in detail elsewhere (Pries et al., 1986; Ley et al., 1986). Briefly, the data used here were obtained from observations of the mesentery of a male Wistar rat. The microvascular network in an area approximately 8 × 11 mm was photographed in a sequence of about 350 overlapping frames. These were assembled in a photomontage, and the network structure was traced on a transparent overlay. Each vessel segment and branch point (node) in the network was labeled, and the topology of the network was expressed in the form of a list segments with associated nodes. Flow direction in each segment was noted. The diameter of each segment was determined manually from the photonegatives, projected at a magnification of 1000 ×. This gave an estimated uncertainty in diameter measurement of about ±0.5 μm.

Hematocrits were estimated in each segment either by cell counting or by a photodensitometric technique (Pries et al., 1983, 1986). The latter technique has been calibrated by measuring the optical density of glass tubes perfused with blood of known discharge hematocrit.

The network chosen for study had the property of being relatively self-contained; that is, the number of segments entering and leaving the network (65) was small relative to the number of segments in the network (913). The values of the flows in the entering and leaving segments are required as "boundary conditions" for the computation. Because the flow rates in these particular segments were not measured directly, values were assigned based on typical flow rates in vessels of similar diameters observed in other rat mesenteric preparations under the same conditions. In about five of these boundary segments, the assumed flow rate was adjusted in the simulation so as to ensure correct flow directions in neighboring segments. Hematocrits in each inflowing segment are also required for the computation, and were assigned values as measured in the network under study.

Rheological Behavior of Blood in Networks

The simulation requires assumptions on the rheological behavior of blood in microvessels. Neither the Fahraeus effect nor the apparent viscosity of blood has been determined directly in rat mesenteric microvessels. The results presented here are based on determinations of blood viscosity and Fahraeus effect in glass tubes with diameters corresponding to the microvessels in the network.

In the simulation, the reduction of tube hematocrit relative to discharge hematocrit due to the Fahraeus effect was expressed as a function of tube diameter and discharge hematocrit, through an empirical relationship based on *in vitro* determinations using human blood (see Chien et al., 1984, p. 231):

$$\frac{H_T}{H_D} = H_D + (1 - H_D)(1 + 1.7e^{-0.415D} - 0.6e^{-0.011D}) \tag{4.1}$$

where D is the tube diameter and H_T and H_D are the tube and discharge hematocrits (see Fig. 4.1). The parameters in the exponents of this equation reflect a scaling of tube diameter with (mean cell volume)$^{1/3}$, to account for the difference in size between rat and human red cells.

The apparent viscosity of blood was similarly expressed as a function of diameter and discharge hematocrit, based on *in vitro* data (see Chien et al., 1984, p. 231):

$$\frac{\eta_{app}}{\eta_{plasma}} = 1 + \frac{e^{H_D\alpha} - 1}{e^{0.45\alpha} - 1}(110e^{-1.424D} + 3 - 3.45e^{-0.035D}) \tag{4.2}$$

$$\text{where} \qquad \alpha = 4(1 - e^{-0.168(D-2.25)})$$

In view of the uncertainty regarding the applicability of the *in vitro* viscosity data, calculations were also carried out using the simpler assumption of constant viscosity in all segments.

Rules for partition of hematocrit at diverging bifurcations were developed on the basis of *in vivo* observations of blood flow in 65 diverging bifurcations in the rat mesentery, as described by Pries et al. (submitted). For each bifurcation, a relationship was developed between ψ, the flow fraction entering one daughter branch, and ϕ, the cell flux fraction entering the branch. It was found that this relationship could be approximated by the following equation:

$$\text{logit } \phi = B \text{ logit } \frac{\psi - X_0}{1 - 2X_0} + A \tag{4.3}$$

$$\text{where} \qquad \text{logit } x = \ln \frac{x}{1 - x}$$

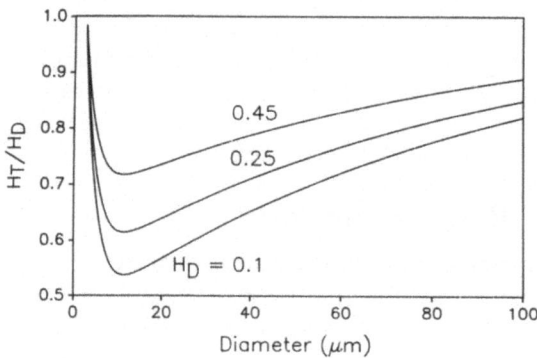

FIGURE 4.1. Variation of Fahraeus effect with tube diameter and discharge hematocrit, based on *in vitro* determinations.

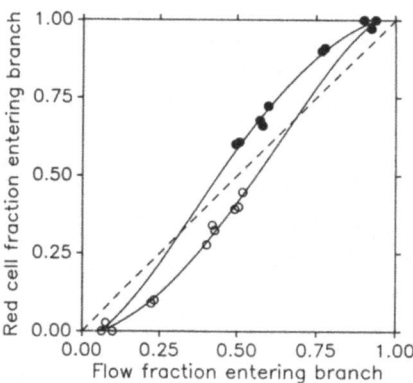

FIGURE 4.2. Example of bifurcation law used in simulation, together with experimental *in vivo* data for one bifurcation. Parent vessel diameter = 7.5 μm; discharge hematocrit = 0.43. Open and closed circles show measured values for the two daughter branches. Solid lines show fitted bifurcation law for the following parameter values: $A = -0.35$, $B = 1.29$, $X_0 = 0.052$.

An example of this law and the fit to the experimental data is given in Figure 4.2. The bifurcation law contains three unknown parameters, A, B, and X_0, which were evaluated for each bifurcation. The parameter A represents the difference between the functional relationships linking ϕ and ψ for the two daughter branches, B is related to the nonlinearity of these relationships, and X_0 is the minimum flow fraction required for any red cells to enter a daughter branch. These parameters were found to vary with H_D and vessel diameters, and the following approximate relationships were developed:

$$A = -6.96 \ln(D_1/D_2)/D$$

$$B = 1 + 6.98 (1 - H_D)/D$$

$$X_0 = 0.4/D$$

where D is the diameter of the parent vessel and D_1 and D_2 are the diameters of daughter vessels. The inverse relationship between all three parameters and D reflects the observation that hematocrit partition becomes more unequal as diameter decreases.

Simulation of Network Flow

The simulations were carried out numerically, using PASCAL language on a personal computer. Some of the underlying concepts have been described by Papenfuss and Gross (1986), although they used a different computational procedure. An iterative method is used here, involving alternate application of two main procedures: (A) computation of pressures at each node and flow in each segment, and (B) computation of the hematocrit in each segment based on the bifurcation law.

The computation of pressures and flows (procedure A) assumes that the hematocrit in each segment is given. The apparent viscosity η_{app} of the segment is then deduced, as described above, and the segment conductance J is given by Poiseuille's law, in the form:

$$J = \frac{\dot{Q}}{\Delta p} = \frac{\pi D^4}{128\, L\eta_{app}} \tag{4.4}$$

where $\dot{Q}$ is flow rate, Δp is pressure drop, and D and L are segment diameter and length. Since the sum of the inflows minus the sum of the outflows at a node must equal 0, the following condition is satisfied at any node at which three segments meet:

$$\dot{Q}_1 + \dot{Q}_2 + \dot{Q}_3 = 0 = J_1(p_1 - p_0) + J_2(p_2 - p_0) + J_3(p_3 - p_0) \tag{4.5}$$

where p_0 is the pressure at the node; p_1, p_2, and p_3 are the pressures at the adjacent nodes; $\dot{Q}_1$, $\dot{Q}_2$, and $\dot{Q}_3$ are the corresponding segment flows (negative for outflows from the node); and J_1, J_2, and J_3 are the corresponding conductances. The "boundary conditions" mentioned above are applied at the single-segment nodes associated with vessels entering or leaving the network. These relationships, when applied at every node in the network, yield a system of linear equations in which the unknowns are the pressures at every node. This system is large but sparse, and is solved efficiently using the iterative method of successive overrelaxation (SOR) (Conte and de Boor, 1981, p. 231).

The assignment of hematocrit in every segment (procedure B) is carried out in terms of discharge hematocrits, and assumes that the segment flows and nodal pressures are given. The principle of conservation of red cell and plasma flow is applied at every node, and the bifurcation law is applied at nodes corresponding to diverging bifurcations. Since the hematocrit in each segment depends on the hematocrit partition at all upstream bifurcations, the hematocrits at a given node cannot be computed until all upstream nodes have been considered. This can be achieved by processing the nodes in order of decreasing pressure. However, this procedure is computationally inefficient because it requires a complete ranking of the nodes by pressure. We developed an alternative algorithm, in which the list of nodes is partially sorted so as to satisfy the condition that any two connected nodes appear in order of decreasing pressure. This ensures that nodes are processed from high to low pressures along every flow pathway.

Procedures A and B are applied alternately ("outer" iteration) until a satisfactory degree of convergence is achieved, typically after 20–30 outer iterations. At each outer iteration, the number of iterations in the SOR ("inner" iterations) was relatively small (20). In the early outer iterations, procedure A then yields very approximate solutions for segment flows, but these are adequate for updating the hematocrits in procedure B. However, since the flows and pressures computed at each outer iteration provide a starting approximation for their computation in the next iteration, these starting approximations become increasingly accurate as the iteration proceeds. Procedure A then converges to high accuracy within 20 inner iterations. Use of the algorithms described above permits simulation of flow in a 913-segment network in several minutes using a personal computer.

Computation of Network Fahraeus Effect

The contributions of vessel Fahraeus effect (VF) and network Fahraeus effect (NF) to total hematocrit reduction (THR) in complete flow cross-sections are expressed by the following relationships (Pries et al., 1986):

$$THR = \frac{\bar{H}_T^A}{\bar{H}_D^{\dot{Q}}} = VF \times NF \tag{4.6}$$

$$\text{where} \qquad VF = \frac{\bar{H}_T^A}{\bar{H}_D^A} \qquad \text{and} \qquad NF = \frac{\bar{H}_D^A}{\bar{H}_D^{\dot{Q}}}$$

The overbar denotes the mean, and the superscripts denote that the mean is weighted by segment cross-section area (A) or by segment flow ($\dot{Q}$). In each case, the mean is taken over a set of vessels constituting a complete flow cross-section as defined above, weighted according to the flow or cross-section area of each segment. We calculated the above parameters for each of a sequence of flow cross-sections containing successively higher branching orders of the diverging microvascular tree. Estimates were made using both measured and predicted hematocrits. To satisfy conservation of red cells and plasma, the flow-weighted mean discharge hematocrit must have the same value in all complete flow cross-sections. In the case of the measured hematocrits, this value was estimated from the measured hematocrits in the most proximal flow cross-sections.

Results

The simulation yields predictions of flow velocities and hematocrits in each segment of the network. Since flow velocities were not measured in the network under study, a direct comparison of predicted and actual flows is not possible. However, the flow direction in every segment was recorded, and the number of segments in which the predicted flow direction is opposite to the observed flow direction provides an indication of the accuracy of the predicted flow pattern. The predictions made assuming a constant viscosity throughout the network showed substantially fewer reversed segments than those made using a viscosity law based on *in vitro* determinations (25 compared to 39). This result was unexpected and is currently being investigated further to evaluate the dependence of the computed flow pattern on the assumed viscosity law.

The principal aim of the study reported here was to test whether the observed hematocrit levels could be accounted for on the basis of the vessel and network Fahraeus effect. In our simulations we therefore assumed constant viscosity in all segments, the condition that gave flow directions corresponding more closely with the observed flows. This condition also results in a flow pattern that is independent of red cell distribution, so that changes in distribution of discharge hematocrit exclusively reflect changes in the hematocrit partition law.

The distribution of discharge hematocrits in the network predicted under these conditions is compared in Figure 4.3 with the distribution of observed values. Good agreement is demonstrated between the means and standard deviations of these distributions. The central peak in the distribution of measured hematocrits is somewhat broader than that of the predicted distri-

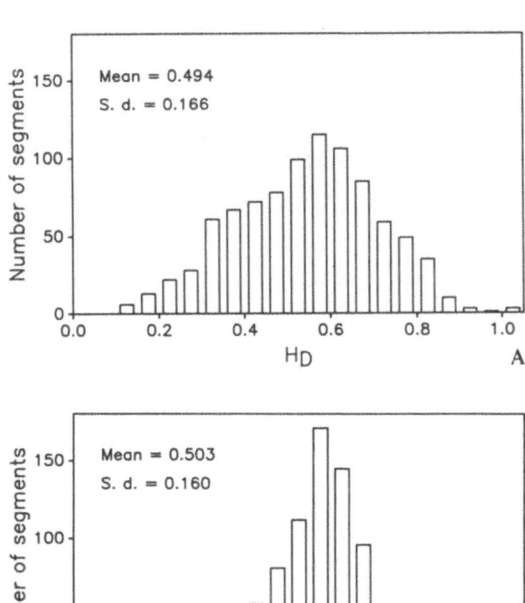

FIGURE 4.3. Distributions of discharge hematocrit H_D. *A*: Measured values. *B*: Predicted values. Mean and standard deviation of H_D are shown for each distribution.

bution. This may result from the uncertainty in the experimental determination of hematocrit (see below). Therefore, in this preparation the observed distribution of hematocrits may be accounted for on the basis of the phase partition at bifurcations as determined *in vivo* in the same tissue.

Since the simulation procedure is based on the actual architecture of the *in vivo* network, segment-by-segment comparisons of hematocrits may be made. Figure 4.4A shows the comparison of predicted and measured discharge hematocrits. A positive correlation between predicted and observed values is obtained, with an r^2 value of .080 and a t value of 8.92, which is significant at the $p < .001$ level. The slope of the relationship between the two variables was computed by bisecting the angle between the two regression lines shown in Figure 4.4. This gave a slope of .979, near to unity. Figure 4.4B shows the corresponding plot including only the first 15 generations of the diverging microvascular tree (see discussion).

Estimates of the network Fahraeus effect in consecutive complete flow cross-sections were obtained from the computed flow and hematocrit distributions (Fig. 4.5B), and compared with estimates based on measured values (Fig. 4.5A). While there is obviously substantial variation between predicted and

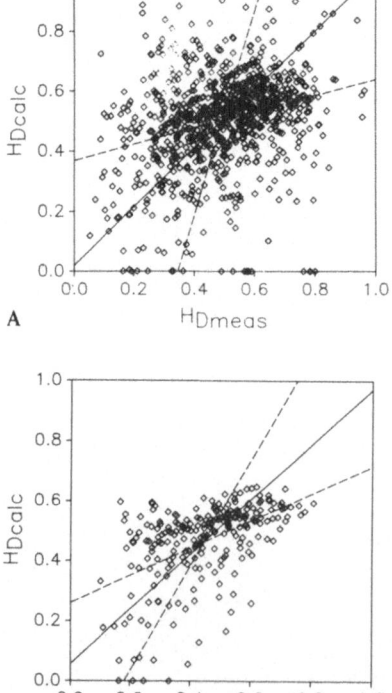

FIGURE 4.4. Comparison of predicted and measured values of H_D. *A*: All segments. *B*: Arteriolar segments to generation 15. Dashed lines give regressions of measured on calculated and calculated on measured values; solid line bisects the angle between the regression lines.

FIGURE 4.5. Network Fahraeus effect (NF), vessel Fahraeus effect (VF), and total hematocrit reduction (THR) based on (*A*) measured and (*B*) predicted values of H_D, as a function of the number of diverging branching generations included in the complete flow cross-section.

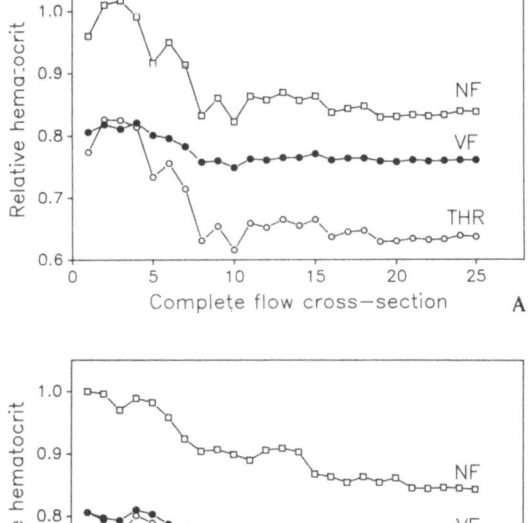

observed values in individual segments (Fig. 4.4), Figure 4.5 shows that the predicted network Fahraeus effect is comparable to the effect estimated based on measured hematocrits.

Discussion

The plots of measured versus predicted discharge hematocrit (Fig. 4.4) show a high degree of scatter. Several possible causes for this variation may be identified. First, there is significant uncertainty in the experimental determination of hematocrit. Comparisons between measured hematocrits in the three vessels meeting at each of a number of bifurcations show that the method has an uncertainty of at least 10–15%. Second, the predicted hematocrit distribution is sensitive to the flow pattern, which is calculated in the simulation. The presence of segments with reversed flow shows that the predicted flow pattern is not accurate. Such inaccuracy is most likely to be due to errors in the estimation of segment conductances, which could result from uncertainties in the estimation of vessel diameter and apparent blood viscosity. Third, the bifurcation law used in the simulation is an approximation to the actual behavior. Even small inaccuracies in the bifurcation law can lead to large cumulative errors as blood flows through many successive diverging bifurcations. Evidence that this effect is responsible for a substantial amount of the observed scatter is provided by Figure 4.4B, which compares measured and predicted hematocrits in all vessel segments with the following property: blood entering the network reaches them along a pathway including up to 14 diverging bifurcations, and no converging bifurcations. The coefficient of determination between measured and predicted hematocrits in this popula-tion is $r^2 = .258$. By comparison, $r^2 = .080$ was obtained for the entire net-work. This suggests that the degree of correlation between observed and measured hematocrits decreases as the number of diverging bifurcations traversed increases.

The comparisons presented above are based on discharge hematocrits. In essence, both photometric and cell counting methods provide measures of tube hematocrit, which is then converted to discharge hematocrit based on *in vitro* determinations of the Fahraeus effect. Desjardins and Duling (1987) have suggested that the Fahraeus effect might be much stronger (i.e., the ratio H_T/H_D might be much smaller) *in vivo* than has been observed *in vitro*. If such an effect were present in the rat mesenteric preparations considered here, then the present results would substantially underestimate the discharge hematocrit levels in the experimental system. According to the results pre-sented above, however, substantially higher discharge hematocrits would be inconsistent with conservation of mass. Based on the estimated tube hemato-crit levels, we conclude that the vessel Fahraeus effect in these preparations does not differ substantially from that observed in glass tubes of corre-sponding diameters.

The presence of a network Fahraeus effect reflects covariance between flow velocity and discharge hematocrit in a complete flow cross-section. This may be expressed quantitatively as follows:

$$NF = \left[1 + \frac{cov^A(v_i, H_{Di})}{\bar{v}^A \bar{H}_D{}^A} \right]^{-1} \qquad (4.7)$$

where the area-weighted covariance of segment velocities and discharge hema-tocrits is defined as

$$cov^A(v_i, H_{Di}) = \sum_i [A_i(v_i - \bar{v}^A)(H_{Di} - \bar{H}_D{}^A)] \bigg/ \sum_i A_i \qquad (4.8)$$

and v_i, H_{Di}, and A_i are respectively the mean bulk velocity, the discharge hematocrit, and the cross-section area in the ith segment.

Figure 4.5 shows that the predicted network Fahraeus effect gives a hemato-crit reduction of about 14% in the final complete flow cross-section, which includes the effects of all diverging bifurcations in the network. The corre-sponding estimate based on measured hematocrits is 16%. This result indicates that a network Fahraeus effect of the observed size, and the underlying correlation between flow velocity and hematocrit, can in fact be explained on the basis of the bifurcation law determined *in vivo*. It is interesting to note that the consistency of simulated and measured network Fahraeus effects implies that the covariance between flow velocity and hematocrit is very similar in the actual and predicted distributions, even though the simulation does not provide good predictions of the hematocrit in individual segments throughout the network.

Conclusions

In conclusion, a method has been developed for detailed simulation of flow patterns and hematocrit distribution in large microvessel networks of arbitrary geometry. This method has been applied to experimental data from the rat mesentery. Predicted hematocrits show a similar distribution to measured hematocrits, although there is substantial variation between predicted and measured values in individual segments. The network Fahraeus effect predicted by the simulation is comparable to that deduced from measured hematocrit values.

Acknowledgment. This work was supported by National Institutes of Health grants HL17421 and HL34555 and the Deutsche Forschungsgemeinschaft.

References

Chien S, Usami S, Skalak, R (1984) Blood flow in small tubes. In Renkin EM, Michel CC (eds) *Handbook of Physiology, Sec 2, The Cardiovascular System.* Vol IV, Pt 1. American Physiological Society, Bethesda, MD, pp 217–249.

Conte SD, de Boor C (1981) *Elementary Numerical Analysis.* McGraw-Hill, New York.

Desjardins C, Duling BR (1987) Microvessel hematocrit: Measurement and implications for capillary oxygen transport. *Am J Physiol* 252:H494–H503.

Fahraeus R (1928) Die Strömungsverhältnisse und die Verteilung der Blutzellen im Gefässsystem. *Klin Wochenschr* 7:100–106.

Johnson PC (1971) Red cell separation in the mesenteric capillary network. *Am J Physiol* 221:99–104.

Klitzman B, Duling BR (1979) Microvascular hematocrit and red cell flow in resting and contracting striated muscle. *Am J Physiol* 237:H481–H490.

Ley K, Pries AR, Gaehtgens P, (1986) Topological structure of rat mesenteric microvessel networks. *Microvasc Res* 32:315–332.

Lipowsky HH, Usami S, Chien S (1980) In vivo measurements of "apparent viscosity" and microvessel hematocrit in the mesentery of the cat. *Microvasc Res* 19:297–319.

Papenfuss H-D, Gross JF (1986) Mathematical simulation of blood flow in microcirculatory networks. In Popel AS, Johnson PC (eds) *Microvascular Networks: Experimental and Theoretical Studies.* Karger, Basel.

Pries AR, Kanzow G, Gaehtgens P (1983) Microphotometric determination of hematocrit in small vessels. *Am J Physiol* 245:H167–H177.

Pries AR, Ley K, Gaehtgens P (1986) Generalization of the Fahraeus principle for microvessel networks. *Am J Physiol* 251:H1324–H1332.

Pries AR, Ley K, Classen M, Gaehtgens P (submitted for publication) Red cell distribution at microvascular bifurcations. *Microvasc Res.*

5
Dispersion of Blood Cell Flow in Microvascular Networks

AXEL R. PRIES and PETER GAEHTGENS

Introduction

It is a generally characteristic feature of microvascular networks that most of the parameters describing the individual vessel segments show markedly heterogeneous distributions. This is certainly true for morphological parameters such as vessel diameter and length, for topological parameters such as segment generation and topological pathway length, and for rheological and hemodynamic parameters such as intravascular pressure, blood volume flow, and local blood cell concentration. On top of these heterogeneities there are considerable fluctuations of the individual parameters in time.

These findings provoke questions about the causes and consequences of the observed heterogeneities. This is of particular relevance for two reasons. On the one hand, heterogeneously perfused networks are less efficient in terms of the exchange between blood and tissue of substances supplied with the blood flow; therefore, any change of the extent of heterogeneity will affect the functional capacity of the network. Alterations of systemic perfusion conditions may, on the other hand, lead to changes in the heterogeneity of the network, since almost all of the above parameters are nonlinearly interrelated.

In principle, it is possible to obtain information on the distributional properties of the microvascular network either by looking at blood flow through a whole organ or tissue or by investigating the elementary processes at the capillary level and then reconstructing the properties of the complete vascular system. We chose the latter approach, with a specific focus on the dispersion of blood cells in microvascular networks.

In this context, there are two potential sources for uneven dispersion of blood particles in microvascular networks (Fig. 5.1): the inherent heterogeneity of blood as a suspension of particles in a continuous phase, and the architectural heterogeneity of the microvascular bed.

Dispersion of blood corpuscles in microvascular networks

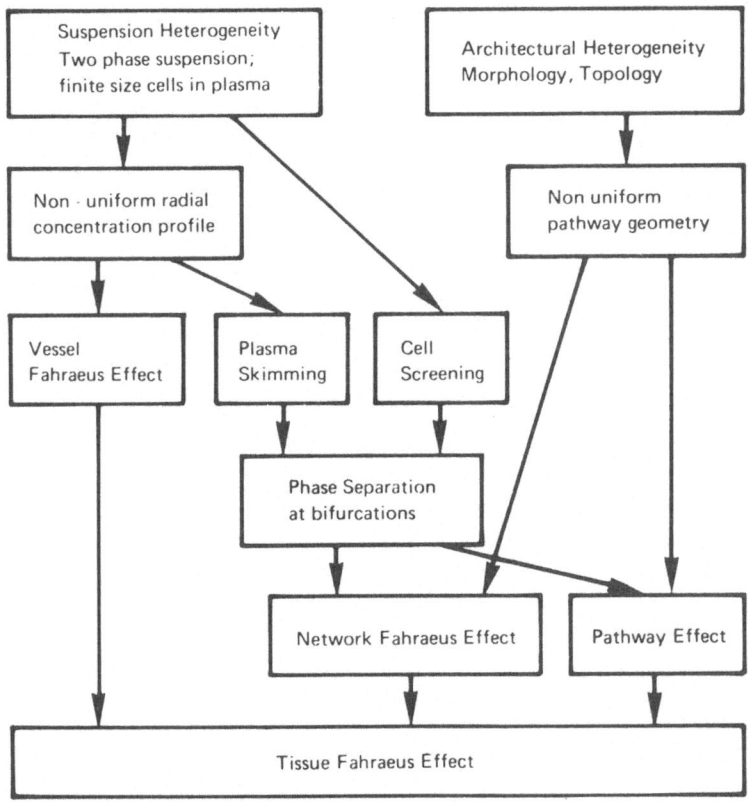

FIGURE 5.1. Block diagram showing the sources and effects of uneven dispersion of blood particles in microvascular networks.

Suspension Heterogeneity

A well-known corollary of the suspension nature of blood is the existence of radial concentration profiles for the different blood cells and corpuscles. Figure 5.2 shows radial red cell concentration profiles determined for arteriolar vessels in the rat mesentery (Pries et al., 1989). These profiles demonstrate that with decreasing vessel diameter there is an increasing difference between the local red cell concentrations in central and peripheral flow regions. Since not only the local hematocrit but also the local flow velocity show a maximum close to the vessel axis, red cells will, on average, travel on faster streamlines than the plasma. Therefore, the mean transit time of red

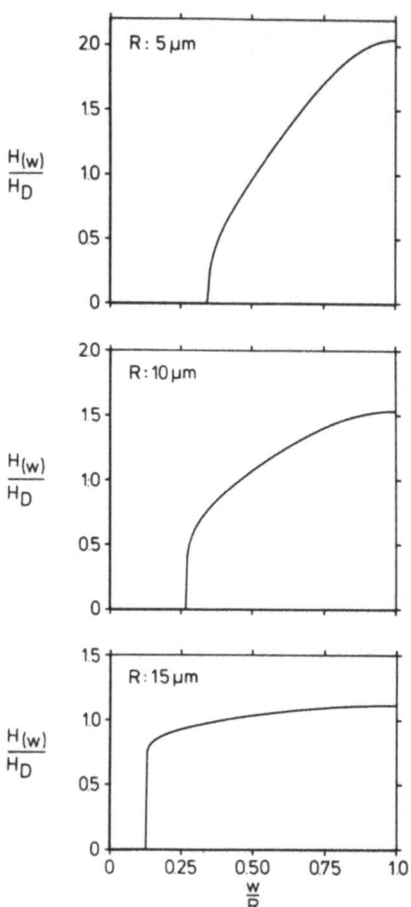

FIGURE 5.2. Radial profiles of local hemato-crit ($H_{(w)}$) normalized with respect to discharge hematocrit (H_D) in arteriolar vessels in the rat mesentery. "w" is the distance from the vessel wall. The results are given for three vessel diameters (10, 20, and 30 μm), the corresponding radius being indicated on the respective panel.

cells through a given vessel segment will be lower than that of the plasma or the blood leading to a reduction in relative red cell concentration in this vessel segment (Fig. 5.3, upper panel). This is called the "Fahraeus effect" and can be described by the equation

$$\frac{v_b}{v_c} = \frac{\bar{t}_c}{\bar{t}_b} = \frac{H_T}{H_D} \tag{5.1}$$

where H_T (the tube hematocrit) is the volume concentration of red cells within a given vessel segment; H_D (the discharge hematocrit) is the volume concentration of red cells in the blood entering or leaving this segment; $\bar{t}_c$ and $\bar{t}_b$ are the mean transit times of cells and blood through the segment; and v_b and v_c are the mean blood velocity and the mean cell velocity in the volume contained in the segment. In the following discussion, this effect will be referred to as

FIGURE 5.3. Definition of vessel (VF), network (NF), and tissue (TF) Fahraeus effect. In all examples, the feeding hematocrit (H_F) and the discharge hematocrit (H_D) are equal. H_T is the volume concentration of red cells within a vessel. v_b and v_c are the average velocities of blood and cells within a vessel. $\bar{t}_b$ and $\bar{t}_c$ are the corresponding average transit times through a vessel or network section. $\overline{H_D}^A$ and $\overline{H_T}^A$ are the area-weighted mean discharge and tube hematocrits on a cross-section through a network, whereas $\overline{H_T}^V$ is the volume-weighted mean hematocrit of all vessels in a given tissue volume.

VESSEL FÅHRAEUS EFFECT

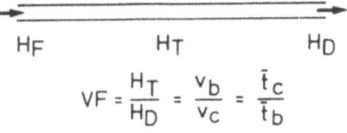

$$VF = \frac{H_T}{H_D} = \frac{v_b}{v_c} = \frac{\bar{t}_c}{\bar{t}_b}$$

NETWORK FÅHRAEUS EFFECT

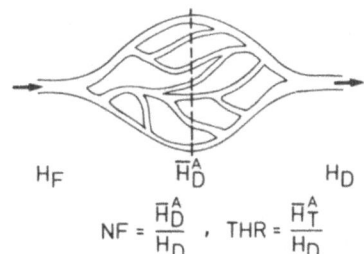

$$NF = \frac{\overline{H_D}^A}{H_D} \quad , \quad THR = \frac{\overline{H_T}^A}{H_D}$$

TISSUE FÅHRAEUS EFFECT

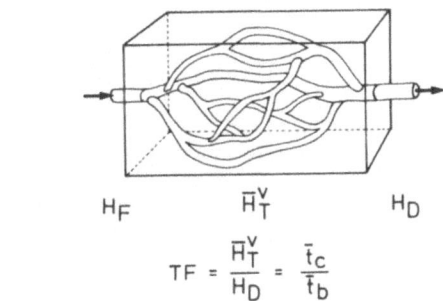

$$TF = \frac{\overline{H_T}^V}{H_D} = \frac{\bar{t}_c}{\bar{t}_b}$$

the *vessel Fahraeus effect* (VF), since it takes place within individual vessel segments.

Since Fahraeus (1928), a number of investigators have determined the magnitude of hematocrit reduction caused by the vessel Fahraeus effect. These studies have shown that hematocrit itself and vessel diameter are the main variables (Chien et al., 1984). Irrespective of quantitative differences between individual studies, it is obvious that the Fahraeus effect does not play a major role in vessels with diameters above several hundred micrometers. It reaches a maximum at a vessel diameter of about 15 μm and decreases at even smaller vessel diameters.

The radial concentration profiles of white cells in tube flow of blood seem to be more complicated than those of red cells since their radial position is strongly influenced by interaction with red cells or red cell aggregates (Fig. 5.4). White cells suspended in plasma tend to travel in flow regions close to

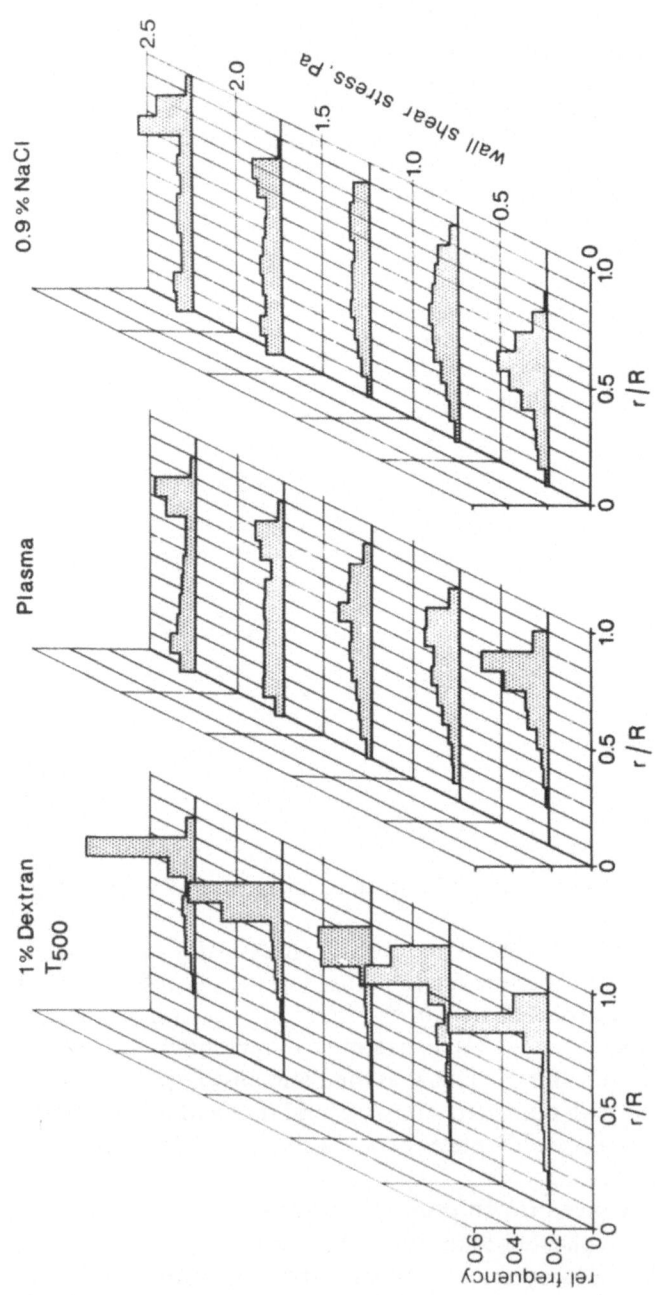

FIGURE 5.4. Radial distribution of white cell concentration measured in a glass tube of 69 μm perfused with blood cells suspended in different media at a hematocrit of 0.4.

the vessel wall, if the flow velocity or wall shear stress values are low. With increased flow velocity more and more white cells are found in central flow regions. If the aggregation tendency of red cells in the suspension is increased, white cells stay marginated over a large range of wall shear stress values, whereas a reduction of red cell aggregation tendency inhibits the strong white cell margination even at low flow rates (Nobis et al., 1982). The radial concentration profile for platelets exhibits the highest values close to the wall, whereas in the central flow regions only a reduced concentration is found (Tangelder, 1982; Tangelder et al., 1986). As a consequence of the radial distributions of white cells and platelets, a vessel Fahraeus effect for these cells can be predicted. For white cells it largely depends on the aggregation tendency of red cells, but on average white cells will travel faster than the blood (if no adhesion at the vessel wall occurs). It is to be expected that platelets travel much slower than the blood does. According to Eq. 5.1, these velocity relations are paralleled by proportional relations between discharge and tube concentrations of the respective particles.

Another consequence of the radial concentration profile of blood cells in microvessels is the so-called *plasma skimming* (Krogh, 1921). At a bifurcation, the daughter branch receiving the lower fraction of volume flow is likely to be fed predominantly from the marginal, low-hematocrit regions of the feeding vessel. Figure 5.5 shows an arteriolar bifurcation in the rat mesentery at two different flow conditions. In the upper panel, both daughter branches receive approximately the same share of the flow in the parent vessel. Under this condition, there is no obvious difference in hematocrit between the two daughter branches. In the lower panel the flow into one daughter branch has been markedly reduced and it is obvious that the hematocrit in this daughter branch is markedly lowered.

In the concept of plasma skimming, the unproportional distribution of red cells and plasma to the daughter branches results from the combination of a hematocrit profile in the feeding vessel and the flow-dividing surface separating the volume flow portions directed into the two daughter branches. In the extreme case, a daughter branch skimming off only the cell-free marginal layer of the parent vessel will receive only cell-free plasma. It has, however, been shown theoretically (Audet and Olbricht, 1986) and experimentally (Pries et al., 1981) that phase separation at bifurcations is not only a result of plasma skimming but might also be evoked by the fluid forces acting on red cells and plasma at the site of the bifurcation, thereby creating an unproportional distribution of red cells and plasma to the daughter branches. This effect, which is independent of a radial hematocrit profile in the feeding vessel, has been called *cell screening* (Cokelet, 1976; Fig. 5.1).

The extent of phase separation at bifurcations is mainly determined by the size of the vessel segments involved and by the hematocrit of the blood fed into the bifurcation (Fenton et al., 1985; Pries et al., 1989). The amount of phase separation increases with decreasing vessel diameter and decreasing feed hematocrit.

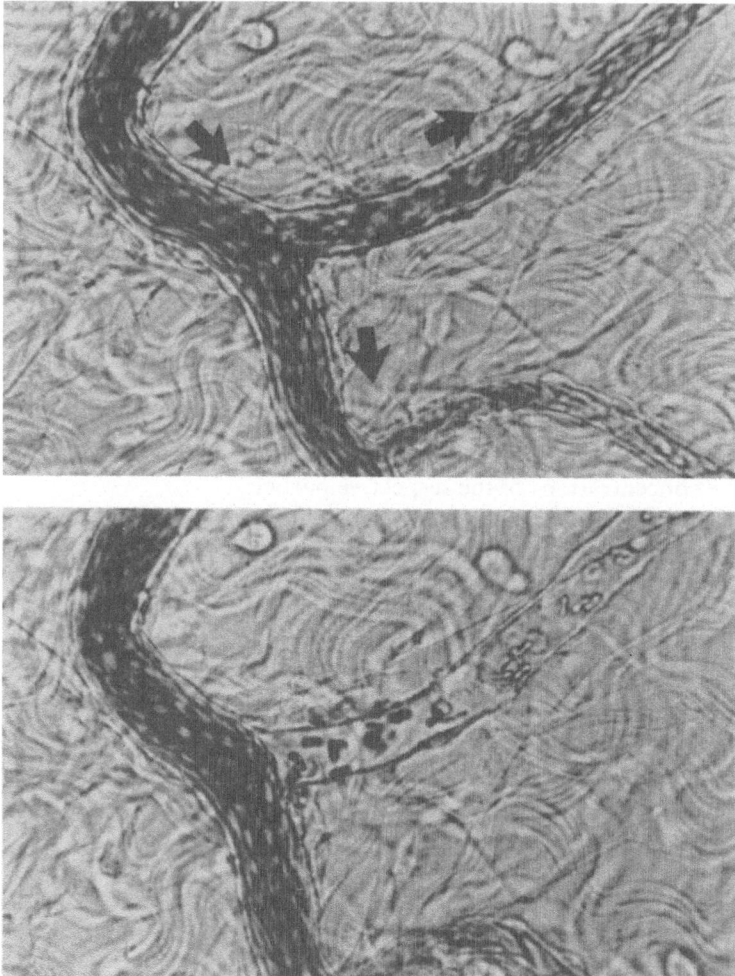

FIGURE 5.5. Arteriolar bifurcation in the rat mesentery. *Upper panel*: Control status (*arrows* show direction of flow). *Lower panel*: Flow into the upper right daughter branch has drastically been reduced by occluding downstream side branches of this vessel.

Architectural Heterogeneity

The relevance of the suspension heterogeneity of blood for cell distribution in a microvessel network strongly depends on network architecture. One of the topological parameters of network architecture is the generation distribution of vessel segments. The generation number of a vessel segment is given by

the number of bifurcations between that vessel and the arteriole feeding
the network (for arteriolar segments) or the draining venule (for venular
segments). Figure 5.6 shows the distribution of arteriolar generation numbers
of terminal segments (capillaries, arteriovenous segments) in the rat mesentery.
From this distribution it can be concluded that a red blood cell might pass
between 2 and 24 bifurcations from the inflow arteriole (diameter about
35 μm) until it reaches a terminal segment. Similar findings are obtained for the
venular tree such that the total number of bifurcations passed by the blood
during transit through the network from the arteriolar inflow to the venular
outflow may vary between 5 and 45.

At each of the successive divergent bifurcations on the individual flow
pathways phase separation effects will take place and result in a higher
hematocrit in the faster flowing daughter branch. This leads to two distinct
phenomena in the particle distribution in the network, which are called here
network Fahraeus effect and *pathway effect* (Fig. 5.1).

As shown in Figure 5.3, the *network Fahraeus effect* (NF) represents a
generalization of the Fahraeus principle to the flow in networks. While the
prerequisite for the vessel Fahraeus effect is a positive correlation between
local hematocrit and velocity on a cross-section of a single vessel, the basis
for the network Fahraeus effect is the correlation between hematocrit and
velocity among the vessel segments of a cross-section through the network
(Pries et al., 1986). If, on average, the faster vessels exhibit the higher hematocrits,
red cells will overproportionally travel on the faster flow pathways. This will
lead to a reduction of red cell transit time compared to the transit time of
blood. The network Fahraeus effect therefore leads to a reduction of average
discharge hematocrit on a flow cross-section through the network ($\overline{H_D^A}$)
compared with the discharge hematocrit in the vessel segment feeding that
network (H_D). The combination of vessel and network Fahraeus effects is
called *total hematocrit reduction* (THR).

Both vessel and network Fahraeus effect increase from proximal to distal
complete flow cross-sections (Fig. 5.7), but for quite different reasons. The
increase of vessel Fahraeus effect is mainly due to the decreasing average
diameter of vessel segments towards the distal flow cross-sections. The network
Fahraeus effect increases, since from proximal to distal flow cross-sections the
correlation between hematocrit and flow velocity increases. Under normal
flow conditions, the vessel Fahraeus effect is by far more relevant than the
network Fahraeus effect, which accounts for about 20% of the total hematocrit
reduction.

The second effect caused by the combination of phase separation at bifurcations and the architectural heterogeneity of networks is the *pathway effect*.
Figure 5.8 shows a very simple terminal network consisting of a feeding
arteriole, a draining venule, and four connecting arteriovenous (AV) vessel
segments. At the first bifurcation the small side branch of the arteriole receives
a lower flow fraction than the continuing main vessel and, because of the phase
separation effects, a reduced hematocrit. This leads, in turn, to an increased

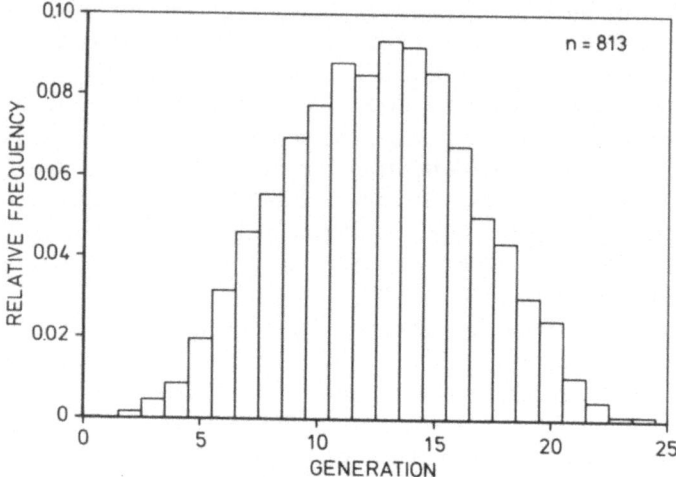

FIGURE 5.6. Frequency histogram of generation numbers of terminal segments determined in arteriolar vessel trees in the rat mesentery.

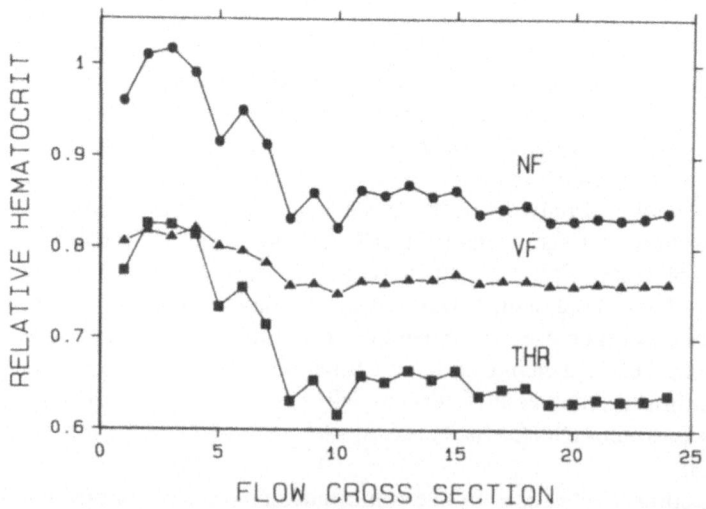

FIGURE 5.7. Hematocrit reduction due to network Fahraeus effect (NF), vessel Fahraeus effect (VF), and total hematocrit reduction (THR) on successive complete flow cross-sections measured in an arteriolar vessel tree in the rat mesentery. The input arteriole of this vessel tree, which represents the complete flow cross-section 1, had a diameter of 35 μm.

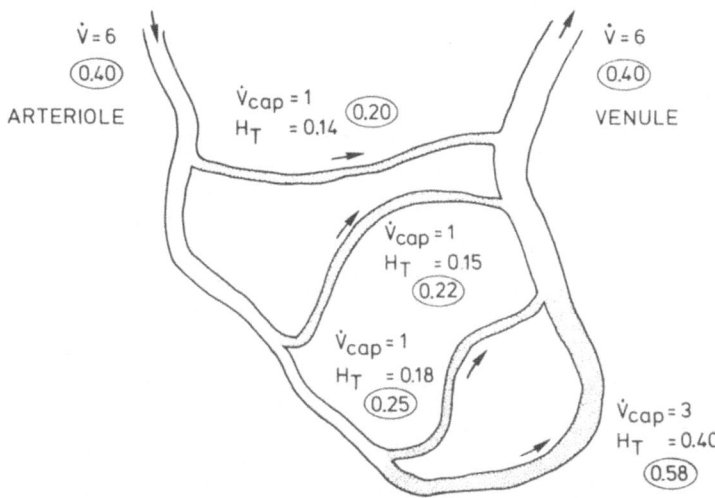

FIGURE 5.8. Simple hypothetical microvascular network consisting of a feeding arteriole, a draining venule, and four connecting arteriovenous vessel segments. Given are volume flow rates ($\dot{V}$), tube hematocrits (H_T), and discharge hematocrits (encircled).

hematocrit in the main channel. This hematocrit increase continues along the flow pathway down to the last AV segment. In a nonsymmetrical network in which AV segments sequentially originate from a continuing main arteriole, red cells will therefore travel on longer flow pathways compared to the plasma or the blood, and this increases red cell transit time compared to that of plasma. Figure 5.9 shows corresponding results for the rat mesentery. Both arterioles and AV segments of high generations exhibit hematocrit levels above the input hematocrit. A similar phenomenon is observed for white cells (Fig. 5.10). Here, the increase of cell concentration with increasing generation number can be suppressed by adding high-molecular-weight dextran to the blood and thereby displacing white blood cells to marginal flow regions of the arterioles because of enhanced red cell aggregation. This leads to a greater number of white cells being skimmed off into the earlier side branches and consequently reduces the accumulation of cells along the arteriolar tree.

Vessel and network Fahraeus effect as well as the pathway effect alter the transit time of cells through a given tissue compared to the transit time of blood. Consequently, they will also change the tissue hematocrit relative to the hematocrit of the supplying blood. For red cells in the mesentery, the vessel and the network Fahraeus effect will decrease transit time and also the tissue hematocrit, whereas the pathway effect will tend to increase these parameters. The combined result of these effects will determine the average hematocrit of the blood contained in all vessels of a tissue element relative to the inflowing

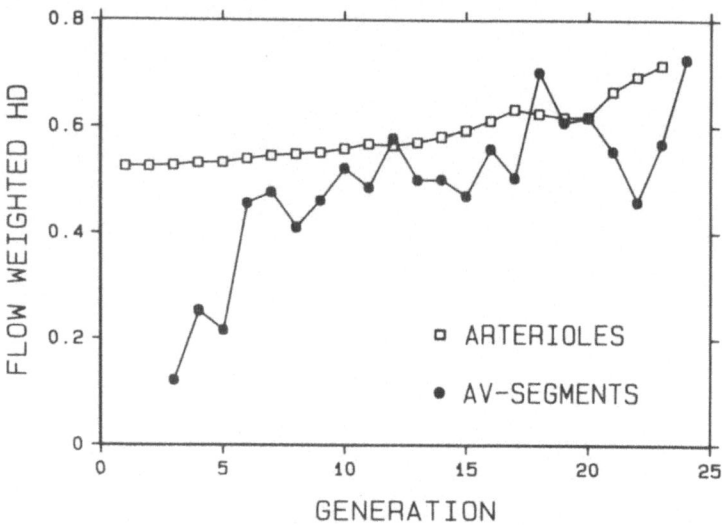

FIGURE 5.9. Flow-weighted mean discharge hematocrits in the arterioles and terminal (AV) segments of consecutive generation levels. Data are given for the same arteriolar vessel tree as in Figure 5.7.

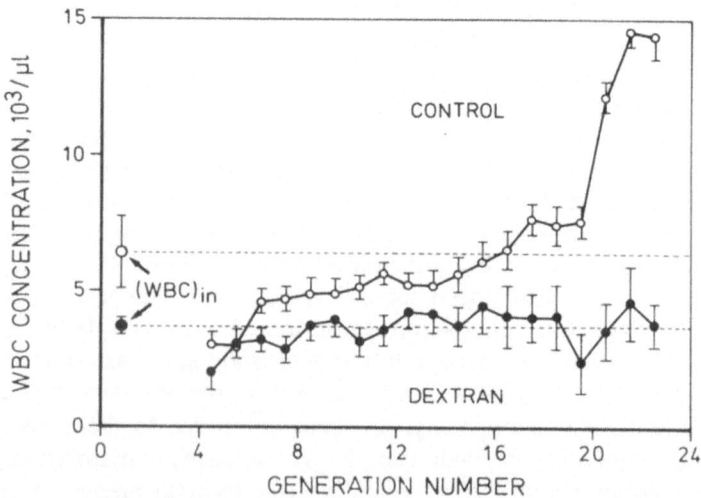

FIGURE 5.10. White cell (WBC) concentration in terminal (AV) segments of the rat mesentery as a function of generation number under control conditions and after administration of high-molecular-weight dextran. In addition to the values on individual generation levels, the systemic white cell concentrations are given (WBC$_{in}$).

hematocrit. This is called here the *tissue Fahraeus effect* (Fig. 5.3). While it is difficult to estimate the actual balance between these three effects from data obtained by intravital microscopy, this is much easier using data from whole organ studies.

Tissue Fahraeus Effect

The Stewart-Hamilton relationship states that the average transit time of blood $(\bar{t}_b)$ through a tissue volume element is equal to the intravascular blood volume (V_b) in this element divided by the blood volume flow rate $(\dot{Q}_b)$ through this element:

$$\bar{t}_b = \frac{V_b}{\dot{Q}_b} \tag{5.2}$$

This relationship can be applied to red cells (or other blood particles) by using the distribution volume of red cells (V_c) and the red cell volume flow rate $(\dot{Q}_c)$ to calculate the average transit time of red cells:

$$\bar{t}_c = \frac{V_c}{\dot{Q}_c} \tag{5.3}$$

The discharge hematocrit of the blood supplying the tissue (H_D) is given by

$$H_D = \frac{\dot{Q}_c}{\dot{Q}_b} \tag{5.4}$$

and the tissue hematocrit (H_{tissue}), which is equal to the volume-weighted tube hematocrit of all vessels in the tissue element $(\overline{H_T}^V)$, is given by

$$H_{tissue} = \overline{H_T}^V = \frac{V_c}{V_b} \tag{5.5}$$

By combining Eqs. 5.2 and 5.3, the relation between the transit times of red cells and blood is obtained as

$$\frac{\bar{t}_c}{\bar{t}_b} = \frac{V_c}{\dot{Q}_c} \times \frac{\dot{Q}_b}{V_b} = \frac{V_c}{V_b} \times \frac{\dot{Q}_b}{\dot{Q}_c} \tag{5.6}$$

Using Eqs. 5.4 and 5.5 this equation can be rewritten as

$$\frac{\bar{t}_c}{\bar{t}_b} = \frac{\overline{H_T}^V}{H_D} \tag{5.7}$$

Eq. 5.7 shows that the ratio between the average transit times of cells and blood is identical to the ratio between the tissue hematocrit and the hematocrit of the blood feeding the tissue (Fig. 5.3, lower panel). Therefore, the combination of vessel Fahraeus effect, network Fahraeus effect, and pathway effect on the cell distribution in the tissue, and thus the tissue Fahraeus effect, can in analogy to the vessel Fahraeus effect be described either by a hematocrit ratio or a ratio of transit times of cells compared to those of blood.

This concept allows the comparison of data on tissue hematocrits with data from indicator dilution studies in which transit times of red cells and blood have been obtained. Figure 5.11 shows the relation between tissue and large vessel hematocrit obtained in a number of tissues by Gibson et al. (1946) and Chien and Simchon (1988). This figure shows that the hematocrit reduction by the tissue Fahraeus effect varies from values below 0.3, indicating that the transit time of red cells is only 30% of the transit time of blood, to values of 2, which means that the transit time of red cells doubles the transit time of blood. Since in the respective studies all organs have been perfused with the same blood, the gross differences must be due to differences in the functional architecture of the respective microvascular beds. This includes the basic morphological features of the vessel tree as well as its effective functional characteristics, which are determined by the distribution and magnitude of vascular tone.

The exact amount of tissue Fahraeus effect is not known for platelets or white cells. In the case of white cells, however, the analysis presented may also be pertinent to the differentiation of "marginated" and "circulating" pools, which essentially differ by their respective transit times.

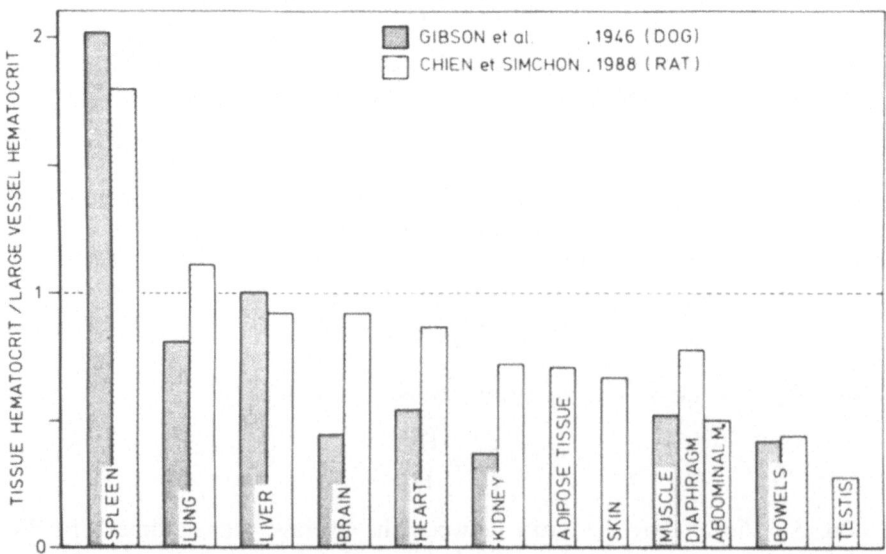

FIGURE 5.11. The tissue Fahraeus effect (tissue hematocrit/large vessel hematocrit) in different tissues.

Summary

The generalization of the Fahraeus concept from a single microvessel to whole organs leads to a new understanding of the rheological, morphological, and physiological mechanisms determining blood cell transit through vascular networks. The implications show that functional properties of such networks are reflected in the transit time distributions of the various blood cells, which may be different due to single particle rheology. Future work will have to demonstrate whether the improvement of the interpretation of transit time distributions will help to identify the physiological mechanisms relating network hemodynamics to network exchange capacity.

References

Audet DM, Olbricht WL (1986) The motion of model cells at capillary bifurcations. *Microvasc Res* 33:377–396.

Chien S, Simchon S (1988) Tissue hematocrit and hemorheology. In Tomita M, Sawada T, Naritomi H, Heiss WD (eds) *Cerebral Hyperemia and Ischemia: From the Standpoint of Cerebral Blood Volume.* Excerpta Medica, Elsevier Science Publ. BV, Amsterdam, pp 39–44.

Chien S, Usami S, Skalak R (1984) Blood flow in small tubes. In Renkin EM, Michel CC (eds) *Handbook of Physiology, Section 2, The Cardiovascular System,* Vol IV, Part 1. American Physiological Society, Bethesda, MD, pp 217–249.

Cokelet GR (1976) Macroscopic rheology and tube flow of human blood. In Grayson J, Zingg W (eds) *Microcirculation,* Vol 1. Plenum Press, New York, pp 9–31.

Fahraeus R (1928) Die Strömungsverhältnisse und die Verteilung der Blutzellen im Gefässsystem. *Klin Wochenschr* 7:100–106.

Fenton BM, Carr RT, Cokelet GR (1985) Non-uniform red cell distribution in 20–100 μm bifurcations. *Microvasc Res* 29:103–126.

Gibson JG, Seligman AM, Peacock WC, Aub JC, Fine J, Evans RD (1946) The distribution of red cells and plasma in large and minute vessels of the normal dog, determined by radioactive isotopes of iron and iodine. *J Clin Invest* 25:848–857.

Krogh A (1921) Studies on the physiology of capillaries. II. The reactions to local stimuli of the blood vessels in the skin and web of the frog. *J Physiol (Lond)* 55:412–422.

Nobis U, Pries AR, Gaehtgens P (1982) Rheological mechanisms contributing to WBC-margination. In Bagge U, Born GVR, Gaehtgens P (eds) *White Blood Cells, Morphology and Rheology as Related to Function.* Martinus Nijhoff, Den Haag, pp 57–65.

Pries AR, Albrecht KH, Gaehtgens P (1981) Model studies on phase separation at a capillary orifice. *Biorheology* 18:355–367.

Pries AR, Ley K, Gaehtgens P (1986) Generalization of the Fahraeus principle for microvessel networks. *Am J Physiol* 251:H1324–H1332.

Pries AR, Ley K, Claassen M, Gaehtgens P (1989) Red cell distribution at microvascular bifurcations. *Microvasc Res.* 38:81–101.

Tangelder GJ (1982) Distribution and orientation of blood platelets flowing in small arterioles. Dissertation, University of Limburg, Maastricht, Netherlands.

Tangelder GJ, Slaaf DW, Muijtjens AMM, Arts T, Oude Egbrink MGA, Reneman RS (1986) Velocity profiles of blood platelets and red blood cells flowing in arterioles of the rabbit mesentery. *Circ Res* 59:505–514.

6
Bioengineering Analysis of Blood Flow in Resting Skeletal Muscle

Geert W. Schmid-Schönbein, Thomas C. Skalak, and Donald W. Sutton

Introduction

Blood flow in skeletal muscle serves to supply nutrients to the muscle fibers and to remove metabolites; it controls tissue growth and atrophy, and determines the response of muscle in disease. The flow rate in the skeletal muscle capillaries can be rapidly adjusted from the low-perfusion conditions of resting muscle to higher flow rates in exercising muscle. Such flow increases are accompanied by shifts in blood cell flux through the microvascular network and in cell metabolism, comprising a highly integrated sequence of events. Skeletal muscle makes up a large part of the body mass and is therefore likely a major determinant of the total peripheral resistance. All of these are reasons we wish to develop a thorough understanding of blood flow in this organ.

In spite of the fact that in the last decades numerous experimental results on blood flow in skeletal muscle were gathered and extensive microvascular observations were made (Hudlická, 1973; Granger et al., 1984), no integrated view of the skeletal muscle circulation has been developed. The main reason for this deficiency is that we are missing a reliable mathematical model that permits analysis of the experimental results. Models are indispensable in modern medical research. They are tools that can be used to unify under one framework the results of widely different experiments. For our purpose, a model will serve to bridge the gap between microscopic and macroscopic observations, and it will serve as a platform from which new questions can be raised. A good mathematical model serves to unify the experiments of the past and should point toward new discoveries in the future.

What are the essential elements that make up a realistic model of blood flow? The answer to this question is not unique for the skeletal muscle circulation but could equally apply to many other organs as well. As a starting point we need:

1. A realistic picture of the vascular and microvascular anatomy, including vessel geometry and branching pattern.
2. A model of the rheological properties of the blood vessel walls, both during passive distention and active contraction.

3. A model of the biophysical properties of blood and the circulatory blood cells, in both the passive and active states.

Once such detailed information is available, mathematical analysis based on biomechanical principles can proceed. We can make predictions and test them by independent experiments. This is our goal. At this stage only a limited data set is available for skeletal muscle, which may serve as background for points 1 through 3. We will therefore focus the discussion in this chapter on the specialized case of resting rat skeletal muscle with dilated blood vessels.

Other approaches to the analysis of skeletal muscle blood flow have been proposed. Bauer et al. (1985) and Braakman (1988) presented theories based on electrical analogs. These analog models serve to describe time-dependent phenomena, and can be used to compute network impedances, but are based on a compartmentalization of the vascular network. Thus, the actual arrangement of the microvascular network is not accounted for, and no independent testing of analog parameters in terms of microcirculatory variables is possible. Network models of the microcirculation in organs other than skeletal muscle have been reviewed by Popel (1987).

A natural starting point for a discussion of organ blood flow is the microcirculation. In this vast network of blood vessels, the Reynolds number is generally small, typically of the order of 10^{-1} to 10^{-3}. The Womersley number is also small, of the order of 10^{-1} to 10^{-2}, so that in the presence of a Newtonian fluid the velocity profiles are fully developed (Atabeck, 1980). Under these conditions convective and inertial forces are negligible compared with the viscous force. Therefore, using index notation, the equation of motion reduces to Stokes' approximation:

$$0 = \frac{\partial p}{\partial x_i} + \mu \frac{\partial^2 v_i}{\partial x_j \partial x_j} \tag{6.1}$$

using Einstein's summation convention for repeated indices. p is the fluid pressure, v_i is the fluid velocity, x_i refers to spatial coordinates, μ is the plasma viscosity, and the subscript i ($=1, 2, 3$) specifies the three orthogonal directions. At physiological pressures, plasma and whole blood are incompressible, so the equation of continuity is

$$\frac{\partial v_i}{\partial x_i} = 0 \tag{6.2}$$

Neglect of the inertia forces is possible if no large arteries are present. In the following we will focus on skeletal muscle in the rat, so that this assumption is satisfied. Skeletal muscle in larger species or in man may have large vessels with inertia, but their numbers are small in comparison with the microvessels where Stokes' approximation is valid.

One of the interesting characteristics of the Stokes equations (6.1 and 6.2) is that the solution is sensitive to the details of the boundary conditions, that is, the geometry of individual blood vessels and the way they interconnect within the microvascular network. This sets up the need for a detailed description of the microvascular anatomy in skeletal muscle.

Vascular Micro- and Macroanatomy

In the past, several models of the microcirculation have been proposed. In the majority of these the microanatomy is only partially represented. Perhaps one of the most celebrated models of the microcirculation is Krogh's model of skeletal muscle (Krogh, 1919). It consists solely of a parallel array of capillaries and has enjoyed enormous popularity. The model is able to show qualitative features of microcirculatory transport; however, it fails in a quantitative approach to organ transport. This is a consequence of the oversimplified geometry of the microvascular network. Using Krogh's model, no distinction can be made between slow or fast, young or old, healthy or diseased muscles. To make progress, new microanatomical models need to be developed.

Models are generally simplifications of reality. The more anatomical complexities of a particular microvascular network we wish to represent in a model, the more network parameters we need. Thus, there is a trade-off between the number of network parameters we select and the number of simplifications we make. Our hope is that if we introduce the minimum number of network parameters necessary to account for all the vessels in the skeletal muscle network, the main features of blood flow are represented. However, refinements with more complexities and consequently more network parameters are possible at any step (Chen, 1983; 1984), and may likely have to be introduced in the future to analyze in detail specific medical problems.

Overview

One of the characteristics of skeletal muscle, even of different animal species, is that it is supplied by several arteries, denoted here as *feeding arterioles*. This feature was recognized more than a century ago in the famous reports on the skeletal muscle circulation by Spalteholz (1888). However, in many modern perfusion studies this fact is neglected, and muscles are often perfused after dissection with only a single feeder intact. In the special case of the rat spinotrapezius muscle, the arterial supply vessels arise from a side branch of the thoracodorsal artery in the anterior region of the muscle and the 11th (and in some rats also the 10th) intercostal artery in the posterior muscle region (Schmid-Schönbein et al., 1986b). These two arteries connect directly inside the muscle via a central vessel, the *arcade bridge arteriole* (Schmid-Schönbein et al., 1987a). The arcade bridge is the largest vessel in the muscle and it is often paired with a venule. From this central arteriole a planar meshwork of arcade arterioles spans out that covers the full extent of the muscle. The arcade system serves to distribute the blood throughout the muscle, and as we will see below, it is able to do so with limited pressure losses. The connection from the arcade arterioles to the capillaries is provided by a set of *transverse arterioles*. Each transverse arteriole has a single root at an arcade arteriole and branches out into an asymmetric bifurcating tree with multiple capillary endings.

The topology of the capillary network is probably the least documented aspect of the circulation, although general features, such as the predominantly parallel orientation between muscle fibers and capillary vessels, have been recognized ever since these networks were first investigated (Spalteholz, 1888). A recent reconstruction of the capillary network in rat skeletal muscle suggests a bundle-like arrangement with capillaries connecting over long distances along the bundle but only a few capillary communications across the bundles. Each bundle has an alternating sequence of arteriolar feeders and venular outflows. The feeders are in fact part of a transverse arteriolar tree. The outflows are denoted as the *collecting venules* (Skalak and Schmid-Schönbein, 1986a). The collecting venules drain into a dense *venular arcade system*, which in turn has multiple outflows to the central veins, denoted as *draining venules*.

For the purpose of blood flow analysis, the connectivity of this network needs to be defined in detail. It is convenient to analyze each hierarchy of blood vessels individually, and in the following discussion we will propose a specific branching schema derived from thin muscles with a planar arcade network, such as rat cremaster, gracilis, or spinotrapezius muscle.

The Arcade Arterioles and Venules

Let us first consider the network of arcade arterioles (Fig. 6.1). They are supplied with blood by a set of feeder vessels, and drained via the transverse arterioles. However, for the moment we will regard the arcade arterioles without connections to the transverse arterioles, as illustrated in Figure 6.2. Such a system of vessels forms a single planar meshwork whose average properties are characterized by the network parameters listed in Table 6.1. The arcade venules form a similar network of vessels, although considerably denser than that of the arterioles. The arcade venules can be characterized by a set of parameters similar to those describing the arcade arterioles, whereas the draining venules form the counterpart to the feeding arterioles.

There are a number of meshwork parameters that can be defined for such a network. In Table 6.1 we list seven parameters that describe important features of its connectivity, for example, the number of arteries or veins connecting from the central circulation to the meshwork, how many vessels and nodes it has, and how many closed arcade loops it has (see Fig. 6.2). Additional parameters can be derived from the quantities listed in Table 6.1. However, only three of these seven branching parameters are independent and need to be measured; the others can be reduced by the following four identities. The total number of nodes in the network is

$$n = n^{\mathrm{I}} + n^{\mathrm{P}} + n^{\mathrm{F}} \tag{6.3}$$

Since the arcades have only dichotomous branches, the number of individual arcade loops is

$$k = \tfrac{1}{2}(n^{\mathrm{P}} + n^{\mathrm{I}}) + 1 \tag{6.4}$$

The number of arcade arterioles (still ignoring the transverse arterioles and

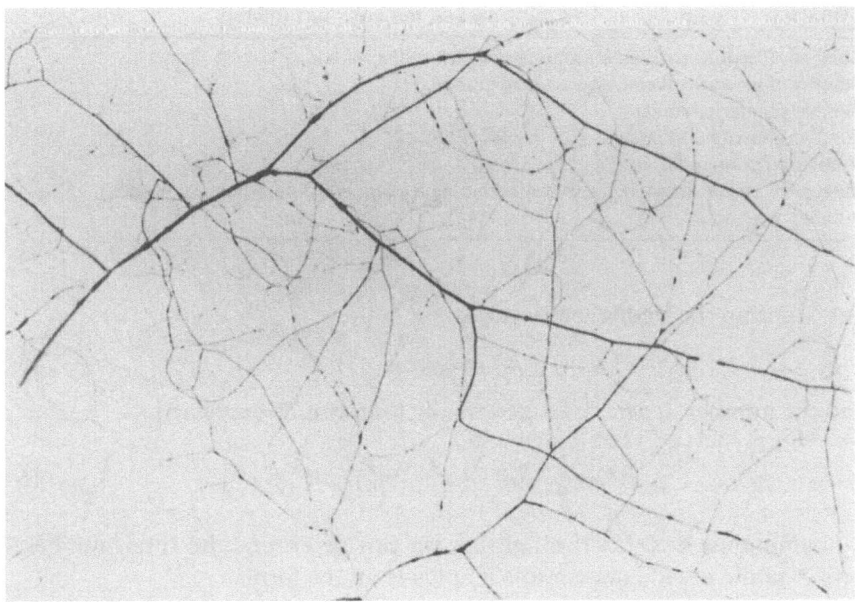

FIGURE 6.1. The arteriolar arcade network in the rat spinotrapezius muscle and underlying latissimus dorsi muscle. The arterioles are filled with a resin material (Batsons, Polyscience, St. Louis) that fills smaller vessels incompletely and thereby serves to highlight the arterioles. Multiple arcade loops with a range of vessel sizes can be seen, and the arcade networks span the entire muscle region. Feeder arterioles are located at left center (a branch from the 11th intercostal artery), bottom right (a branch from the thoracodorsal artery), and bottom center (a branch from the 10th intercostal artery supplying the latissimus dorsi muscle arcades).

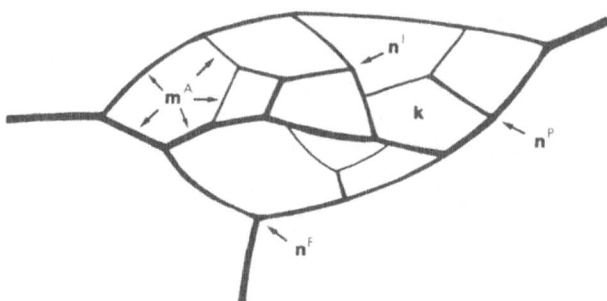

FIGURE 6.2. A schematic of the arteriolar or venular network. n^F represents feeder vessels, n^P is a perimeter node, n^I is an interior node, k is the number of arcade loops, and m_A is the number of vessel segments comprising a single arcade loop.

TABLE 6.1.   Arteriolar and venular arcade network parameters

Number of feeding arterioles (draining venules) to the muscle	n^F
Number of arterioles (venules) per arcade loop	m^A
Number of internal nodes	n^I
Number of peripheral nodes	n^P
Total number of nodes	n
Number of arcade arterioles (venules) excluding feeding arterioles (draining venules)	n^A
Number of arcades	k

not counting the feeder arterioles) is

$$n^A = \tfrac{1}{2}(3n - n^F) \tag{6.5}$$

and the number of arterioles per arcade loop can be expressed as:

$$m^A = \frac{1}{k}[2(n^A - n^F - n^R) + n^F + n^R] \tag{6.6}$$

By combining Eqs. 6.3 through 6.5 we can determine the total number of vessels in the arcade network as n^A plus n^F in the form

$$n^F + n^A = 2n + 1 - k$$

After substitution of Eq. 6.5 into Eq. 6.6 and using the above expression for k, we find

$$m^A = \frac{2(3n - 2n^F - n^P)}{(n - n^F + 2)}$$

For a constant number of feeder vessels, n^F, and increasing n, $3n - 2n^F - n^P \rightarrow 3n$ and $n - n^F + 2 \rightarrow n$. Thus:

$$\lim_{n \to \infty} m^A = 6 \tag{6.7}$$

This result shows that for a large planar dichotomous arcade network, on average a hexagonal pattern is approached. Direct measurements show that in vivo values for m^A fall between 5 and 6.

The network parameters listed in Table 6.1 can be normalized by the muscle volume so that comparisons between muscles are possible. Each parameter has been measured in the spinotrapezius muscle together with the vessel length and diameter distributions (Engelson et al., 1985a, 1985b; Schmid-Schönbein et al., 1987a). Popel et al. (1988) have proposed an alternative form of network description of the arcade arterioles that makes use of an ordering scheme for individual nodes based on their topological position within the arcade meshwork.

The Transverse Arterioles and Their Venular Counterparts

In light of their immediate precapillary position, the transverse arterioles have also been designated as terminal arterioles (Lindbom and Arfors, 1985). They have a dichotomous treelike branching pattern that is similar to that of the collecting venules, with a single root at the arcades and multiple capillary connections (Fig. 6.3). The transverse arteriolar tree may feed not only into a single capillary bundle element, but into several bundle elements either along the same capillary bundle or into neighboring bundles. A similar branching pattern is found for the collecting venules.

The transverse arterioles have a localized strong influence over the blood flow in the capillary network into which they feed. Their diameter is small enough to be able to close the lumen and completely stop blood flow during smooth muscle contraction by deformation of the endothelial cells (Schmid-Schönbein and Murakami, 1985). Their average spacing, z, along the arcades

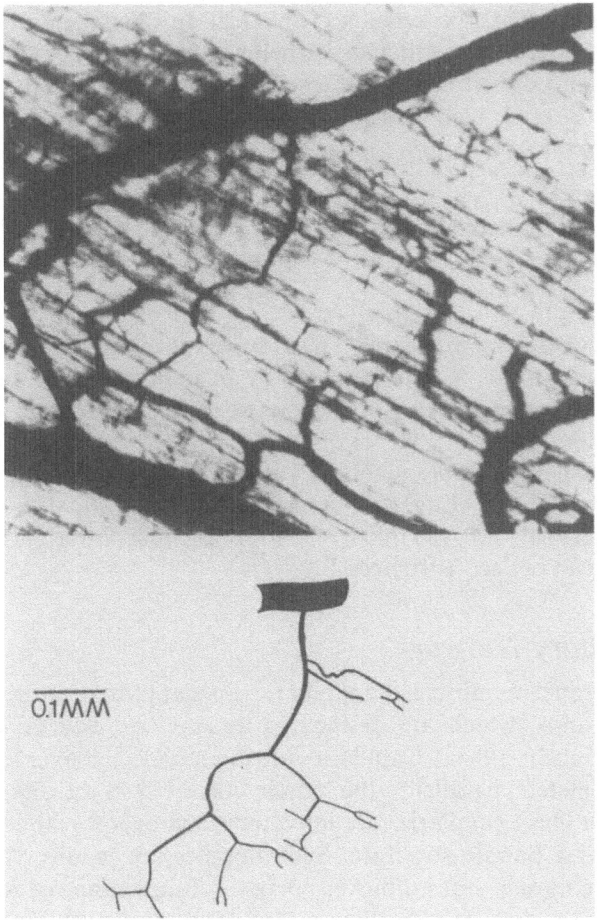

FIGURE 6.3. *Top*: Micrograph of a transverse arteriole in rat spinotrapezius muscle after filling with carbon suspension. *Bottom*: Schematic tracing of the same arteriole, showing the single root at the arcade arteriole, the typical asymmetrical branching pattern, and multiple capillary connections. (From Engelson et al., 1985b.)

TABLE 6.2.   Transverse arteriolar (TA) and collecting venular (CV) network parameters

Average spacing between TA (CV) along the arcade arterioles	z
Branching ratio	R_B
Number of vessels in the first order	N_1
Highest branching order	I
Length branching ratio	R_L
Average length of second-order vessels	L_2
Diameter branching ratio	R_D
Average diameter of second-order vessels	D_2

is considerably shorter on the venous side than on the arteriolar side. The branching pattern for both types of trees is conveniently described by the Strahler tree classification schema (Table 6.2). In this schema the capillaries are designated as the first order, and whenever two vessels of equal order join the next higher order is formed, but when unequal orders join the higher order is carried over to the next branch. The average number of vessels, N_n, in the order n is given by

$$N_n = R_B^{1-n} N_1 \tag{6.8}$$

where R_B is the branching ratio

$$R_B = \frac{1}{I-1} \sum_{i=1}^{I-1} \frac{N_i}{N_{i+1}} \tag{6.9}$$

and I is the highest over in the tree. The average vessel length, L_n, and diameter, D_n, in each order n can be approximated by an analogous schema, except that the length and diameter branching ratio are based on the second order instead of the first order.

Each of the network parameters in Table 6.2 has been measured in the rat spinotrapezius muscle (Engelson et al. 1985a, 1986) and the cat sartorius muscle (Koller et al., 1987) so that a realistic branching pattern of the arterioles and venules can be reconstructed.

The Capillary Network

In the spinotrapezius muscle, the capillaries appear to be arranged in the form of separate units, which are designated as *capillary bundles* (Skalak and Schmid-Schönbein, 1986a). Capillaries are interconnected over long distances (several millimeters) parallel to the muscle fibers, but in the direction normal to the muscle fibers capillaries are interconnected only for limited distances, giving rise to a bundle structure. Such bundles are readily visible in thin regions of the muscle, but in thicker portions of the muscle or regions where the muscle tapers their arrangement needs further clarification. The transverse arterioles and collecting venules feed and drain, in an alternating squence, the capillary bundles, so that an arteriole is always opposed by two venules and vice versa (Fig. 6.4). The arteriole and its opposing venule and their inter-

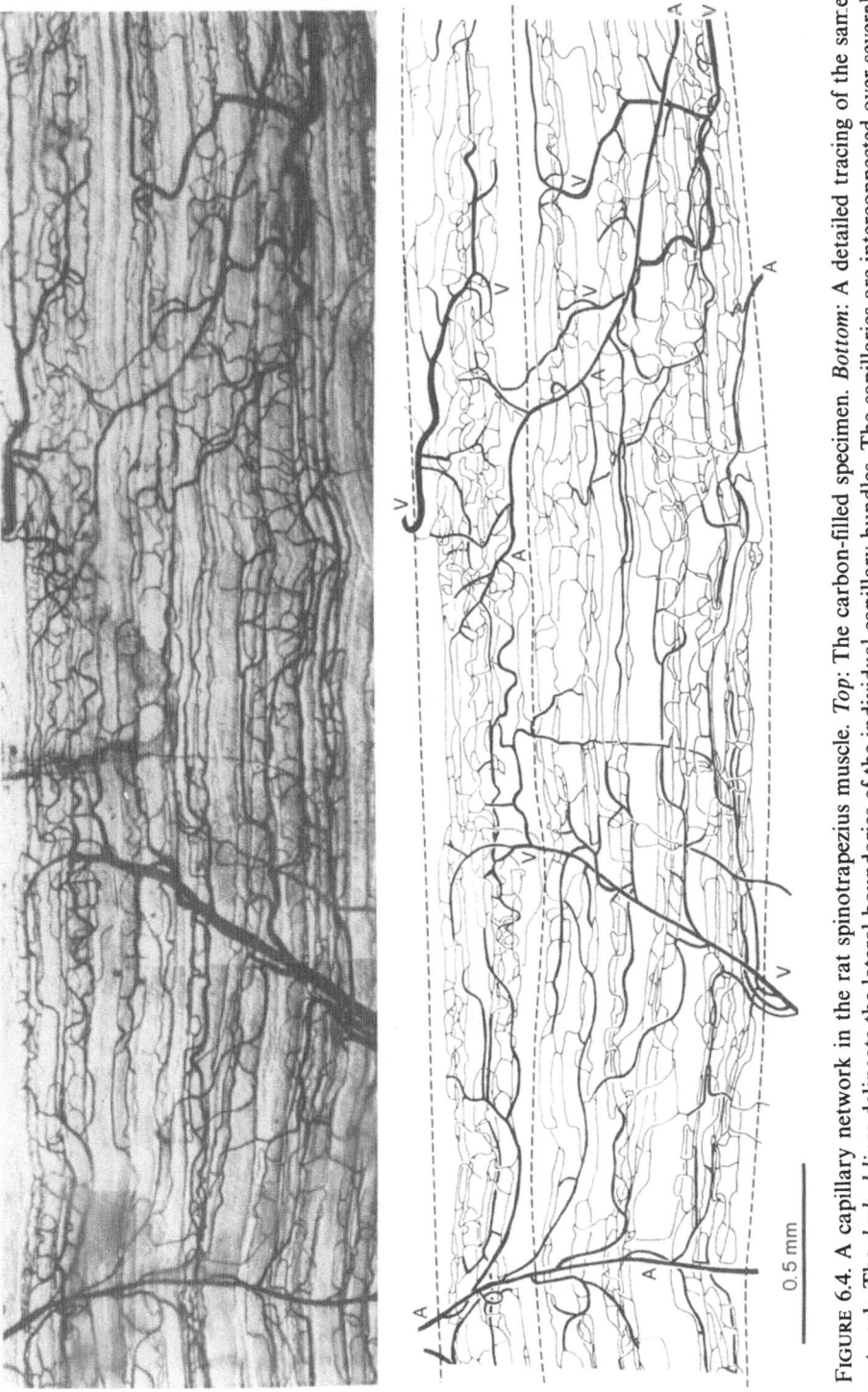

FIGURE 6.4. A capillary network in the rat spinotrapezius muscle. *Top*: The carbon-filled specimen. *Bottom*: A detailed tracing of the same network. The *dashed lines* delineate the lateral boundaries of the individual capillary bundles. The capillaries are interconnected over several millimeters along the muscle fibers. Note the repeated polarity of transverse arterioles (A) and collecting venules (V) along the individual capillary bundles. (From *Microvascular Networks: Theoretical and Experimental Studies*, Schmid-Schönbein et al., 1986a. Figure 4, pg. 44, S. Karger AG.)

Table 6.3. Capillary network parameters

Average number of parallel capillaries per bundle	P
Number of capillary connections on arteriolar side	A
Number of capillary connections on venular side	V
Number of capillary nodes between arteriole and venule	N
Number of capillary back-connections across the arteriole	B_A
Number of capillary back-connections across the venule	B_V
Number of capillary connections between neighboring bundle elements	C

mediate capillary network may be regarded as the building block for a capillary bundle. Accordingly we have denoted it as a *capillary bundle element*.

Table 6.3 contains a list of seven network parameters that represent the minimum number to account for all of the vessels in a capillary bundle element of rat muscle. If fewer parameters are selected, some vessels remain unclassified. Additional parameters can be introduced if more details about the branching pattern are desired. Since the capillary bundles appear in different sizes, it is useful to normalize the branching parameters in Table 6.3 so that comparisons of different capillary bundles and between different muscles can be made. This leads to six nondimensional groups:

$$R_1 = \frac{A}{P}, \qquad R_2 = \frac{V}{P}, \qquad R_3 = \frac{N}{P}$$

$$R_4 = \frac{B_A}{A}, \qquad R_5 = \frac{B_V}{V}, \qquad R_5 = \frac{C}{P}$$

(6.10)

The parameter R_6 is small since the bundles are selected so that capillary communications between bundles are kept to a minimum. Actual values are listed in Skalak and Schmid-Schönbein (1986a). The capillary length and diameter distributions have typical features found for many microscopic blood vessels. The length has a log normal distribution, and the diameter can be closely approximated by a Gaussian distribution with relatively small standard deviation (Skalak and Schmid-Schönbein, 1986a).

Microvessel Wall Properties

Blood vessels exhibit viscoelastic properties, and microvessels are no exception. One of the ways the wall properties of microvessels can be tested is by local inflation at different pressures. If plasma is used, such a procedure causes distention of blood vessels, but the wall will filter fluid at the same time (Lee et al., 1987). In order to uncouple filtration and distention, we filled microvessels in skeletal muscle with a water-immiscible silicone polymer and occluded the venous outflow, thereby establishing a uniform hydrostatic pressure within the microvasculature. The microvessels were then inflated to various pressures while the vessel diameter, d, was measured by intravital microscopy and related to the transmural pressure $P(t) = p(t) - p_t$, where $p(t)$ is the intravascular pressure and p_t the adjacent tissue pressure. The experimental results show that the pressure $P(t)$ at any instant of time is a linear function of the diameter strain (Skalak and Schmid-Schönbein, 1986b).

$$E(t) = \tfrac{1}{2}(\lambda^2 - 1) \tag{6.11}$$

where $\lambda(t) = d(t)/d_0$ and d_0 is a reference diameter for $P = 0$. In the presence of the silicone polymer, such a reference diameter is readily detectable and reproducible. However, with plasma some microvessels, such as the capillaries, tend to collapse if the transmural pressure is kept at zero for prolonged periods of time, since the fluid can escape across the wall. In this case, the reference diameter d_0 has to be established at a finite pressure.

Experimental results with different types of pressure histories can be closely approximated by a model that is analogous to the standard linear solid in viscoelasticity (Fig. 6.5). The governing differential equation is:

$$P + \frac{\beta}{\alpha_1}\dot{P} = \alpha_2 E + \beta\left(1 + \frac{\alpha_2}{\alpha_1}\right)\dot{E} \tag{6.12}$$

where $\dot{P} = dP/dt$ and $\dot{E} = dE/dt$. α_1, α_2, and β are constant coefficients that depend on the properties of the vessel wall. In order to find the diameter for a given pressure history, we need to integrate Eq. 6.12. The advantage of using Eq. 6.12 is that it is a linear differential equation that can be integrated by the elegant integral techniques. Specifically, for an entire pressure history $P(t)$, the diameter strain $E(t)$ can be expressed in the form of a hereditary integral:

$$E(t) = \int_{-\infty}^{t} J(t - \tau)\frac{dP}{d\tau}\, d\tau \tag{6.13}$$

where $J(t)$ is the creep function

$$J(t) = \frac{1}{2}\left(\frac{1}{\alpha_2} - \left(\frac{1}{\alpha_2} - \frac{1}{\alpha_1 + \alpha_2}\right)\exp\left[-\frac{\alpha_1 \alpha_2 t}{\beta(\alpha_1 + \alpha_2)}\right]\right) \tag{6.14}$$

On the other hand, if the diameter strain history $E(t)$ is given, the pressure is

$$P(t) = \int_{-\infty}^{t} G(t - \tau)\frac{dE}{d\tau}\, d\tau \tag{6.15}$$

where $G(t)$ is the relaxation function

$$G(t) = \frac{1}{2}\left(\alpha_2 + \alpha_1 \exp\left[-\frac{\alpha_1}{\beta}t\right]\right) \tag{6.16}$$

Specific values of α_1, α_2, and β have been measured for various hierarchies of blood vessels in rat skeletal muscle (Skalak and Schmid-Schönbein, 1986b).

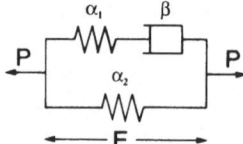

FIGURE 6.5. The standard viscoelastic solid model for the pressure-diameter relationship in skeletal muscle. P is the transmural pressure, E the diameter strain (Eq. 6.11), α_1 and α_2 elastic coefficients, and β the viscous component.

The above model describes the passive properties of blood vessels, and is similar to a viscoelastic model developed by Bauer et al. (1985) for larger arteries in the rat. A review of previous microvessel distensibility measurements was provided elsewhere (Fung, 1978; Skalak and Schmid-Schönbein, 1986b). An interesting outcome of the measurements was the observation that the venules inside the muscle are stiffer than the arterioles, in contrast to other organs with isolated superficial venules, such as the skin, where venules and veins are the most distensible segment of the vasculature.

Blood Flow in Single Vessels

As a starting point, consider the steady motion of plasma with viscosity μ in a skeletal muscle microvessel. Integration of Eqs. 6.1 and 6.2 over a circular cross-section gives Poiseuille's formula relating the flow rate, $\dot{Q}$, to the axial pressure gradient, $\partial p/\partial z$, according to:

$$\dot{Q} = \frac{\pi}{8} \frac{a^4}{\mu} \left(-\frac{\partial p}{\partial z} \right) \qquad (6.17)$$

where $a = d/2$ is the vessel radius. At steady state, the wall distensibility reduces according to Eq. 6.12 to the relationship

$$P = \alpha_2 E \qquad (6.18)$$

At atmospheric tissue pressure the transmural pressure P is equal to p, the intravascular pressure. Substitution of Eqs. 6.11 and 6.18 into Eq. 6.17 gives, after integration over the length, l, of the vessel, the following equation for the flow rate

$$\dot{Q} = \frac{\pi a_0^4 \alpha_2}{48 \mu l} \left[\left(1 + \frac{2p_A}{\alpha_2} \right)^3 - \left(1 + \frac{2p_V}{\alpha_2} \right)^3 \right] \qquad (6.19)$$

where p_A and p_V are the pressures at $z = 0$ and 1, respectively, and $a_0 = d_0/2$ is the reference radius. This is the basic pressure-flow relationship of skeletal muscle. The flow is a function of both the arterial and venous pressure. For constant venous pressure, the flow is a third-order function of the arterial pressure. This prediction can be independently tested by whole organ pressure-flow studies, since if each vessel in the network follows such a third-power relationship, the entire network will follow it as well. Sutton

(1987) has carried out such a study on the rat gracilis muscle. In these experiments the muscle is initially perfused with a Newtonian electrolyte-albumin solution as a control. In each animal the third-power prediction, according to Eq. 6.19, was found to approximate the experimental readings within the measurement error (Fig. 6.6). Furthermore, fixation of the micro-vessels with glutaraldehyde, to transform viscoelastic blood vessels into rela-tively stiff elastic tubes by protein cross-linking, reduces the third-power relationship to an almost linear pressure-flow curve, as expected for rigid tubes.

With the addition of blood cells, the apparent viscosity of the blood, μ_a, is increased over that of plasma, and becomes shear rate dependent. The apparent viscosity is a function of blood cell mechanical properties, aggrega-tion, and adhesion to the endothelium (see review by Skalak et al., 1986). A specific empirical formula for the apparent viscosity, μ_a, which accounts for the non-Newtonian properties is (Schmid-Schönbein, 1988)

$$\mu_a = \left(k_1 + \frac{k_2}{(V_m/d)^\alpha} \right)^2 \qquad (6.20)$$

where V_m is the average velocity and k_1, k_2, and α are empirical coefficients. For $\alpha = 1/2$ (Lipowsky, 1975), substitution into Eq. 6.17 gives

$$\frac{dp}{dz} = -\dot{Q}\frac{8}{\pi a_0{}^4 P'^2}\left[k_1 + k_2\left(\frac{2\pi a_0{}^3}{\dot{Q}}P'^{3/2}\right)^{1/2} \right]^2 \qquad (6.21)$$

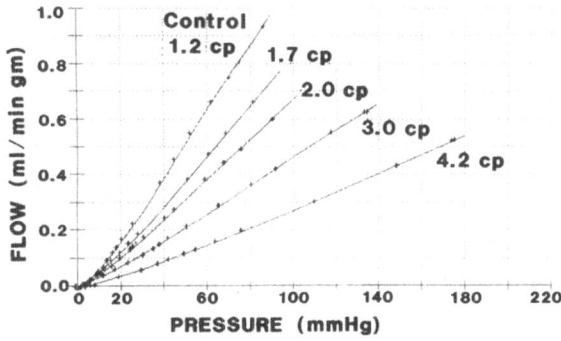

FIGURE 6.6. Experimental pressure-flow curves in the hemodynamically isolated rat gracilis muscle. The muscle is maintained in situ, and the microvasculature is dilated with papaverine and perfused with a Newtonian electrolyte-albumin solution ac-cording to Sutton (1987). Dextran (70,000 MW) was added to increase the viscosity. The arterial pressure is raised while the venous pressure is kept at zero. The experimen-tal error is equal to or less than the size of the symbols. Each line represents a third-order curve fit according to Eq. 6.19. The nonlinearity at lower pressures is solely due to elasticity of the blood vessels and is small due to the relative rigidity of the skeletal muscle microvessels.

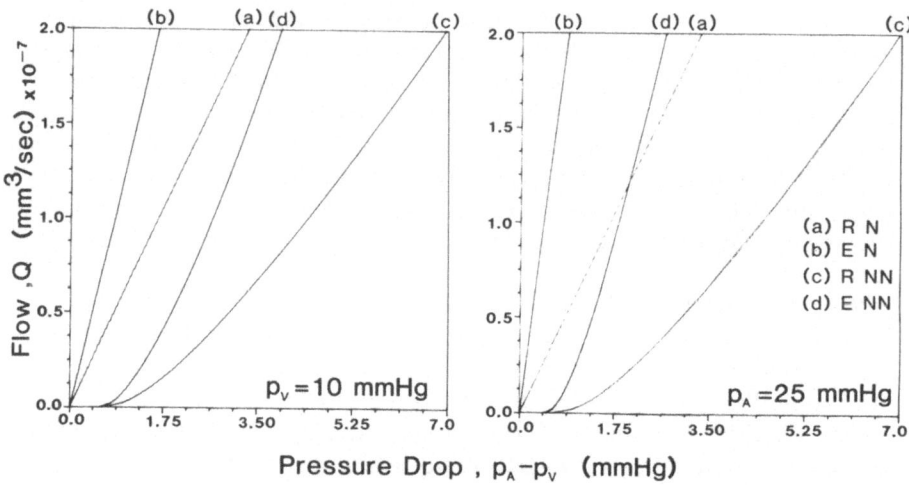

FIGURE 6.7. Theoretical pressure-flow curves for a single microvessel in skeletal muscle (curve (d)) with fixed venous pressure p_V (left) and fixed arterial pressure p_A (right). Curve (a) shows for comparison the linear relationship for a rigid vessel with plasma as a Newtonian fluid (R N), and curve (b) shows the increased flow achieved in an elastic vessel with a Newtonian fluid (E N). Curve (c) shows the decrease in flow due to the addition of red blood cells and presence of non-Newtonian fluid properties (R NN), and curve (d) illustrates the compensatory effect on the flow due to vessel distensibility (E NN). In curve (d), representing skeletal muscle blood vessels, the flow is returned to values near those of curve (a) despite the increase in apparent viscosity caused by the red cells. The flow compensation due to vessel elasticity is most marked for the case with higher mean pressure at right, because of the increased vessel distention. (Reprinted with permission from *Transactions of the ASME, Journal of Biomechanical Engineering*, Schmid-Schönbein, 1988)

with $P' = (1 + 2p/\alpha_2)$. This is the governing equation for an elastic tube filled with a non-Newtonian blood. The equation is readily integrated by use of a Runge-Kutta technique. Figure 6.7 shows such a result. Whereas the red blood cells cause an increase of resistance in the blood vessel, compared with plasma, the vessel distensibility decreases the resistance due to the vessel expansion. Even a mild distention can be effective to compensate for the added resistance caused by the red cells.

So far we have discussed steady blood flow in a single vessel. Next, we wish to discuss blood flow in a skeletal muscle microvascular network.

Blood Flow in the Microvascular Network

Analysis of blood flow throughout the network is useful at several levels. First, results from such a model may be used to examine the detailed dependence of local pressure, flow, and filtration on blood rheology, vessel wall properties, and permeability. Second, such an analysis forms the bridge between micro- and macrocirculation. The distribution of flow in a whole tissue may be examined while precise control of parameters affecting flow is maintained.

Network model formulations may vary in their treatment of vessel network geometry, blood rheology, and vessel wall mechanics. We will construct here a network model, representing the microvasculature in a realistic fashion, and considering the organ with its entire network so that comparison with whole organ pressure-flow experiments is possible. In this section we describe the formulation of the model, some of its predictions, and some verifications by independent in vivo experiment results. Further, we illustrate the utility of the model by application to selected problems in skeletal muscle microcirculation that have been difficult to resolve by experiments only.

Formulation of the Network Model

Network Structure

The basic building block for the skeletal muscle model is the capillary bundle element, that is, the group of capillaries fed and drained by a single arteriole-venule pair as described in "The Capillary Network," above. The number of parallel capillaries in a typical capillary bundle is specified according to the measurements; in the current case it was 30, which is typical for rat skeletal muscle (Skalak and Schmid-Schönbein, 1986a). This allows computation of the other capillary network parameters A, V, N, B_V, and B_A from the previously measured ratios R_1-R_6 (Eq. 6.10). The values of A and V specify the number of first-order arteriolar and venular connections, respectively, and they serve as the order 1 vessels in the Strahler scheme for construction of the model transverse arteriole and collecting venule. Using the measured branch ratios, R_B, for these vessels, the number of vessels in each order of the transverse arteriolar and collecting venular tree is determined and used to construct these trees according to Eqs. 6.8 and 6.9. In the capillary network, the proper number of cross-connecting vessels, $N/2$, and back-connections, B_V and B_A, are then inserted. Vessel lengths and diameters are assigned according to the measured log normal and Gaussian distributions for capillaries (Skalak and Schmid-Schönbein, 1986a), transverse arterioles (Engelson et al., 1986), and collecting venules (Engelson et al., 1985a). The resulting capillary bundle element (Fig. 6.8) contains 260 nodes (vessel bifurcations), and represents an average capillary network in rat skeletal muscle.

Now consider the construction of a single capillary bundle. For this purpose the capillary bundle element of Figure 6.8 is iterated n times, and the n bundle elements are connected together end to end by way of the open capillary ends at the boundaries of each bundle element. This procedure serves to interconnect the capillary bundle elements into a capillary bundle that has repeated transverse arterioles and collecting venules along its length (Fig. 6.9). At the two open endings of the resulting capillary bundle, the capillary ends are

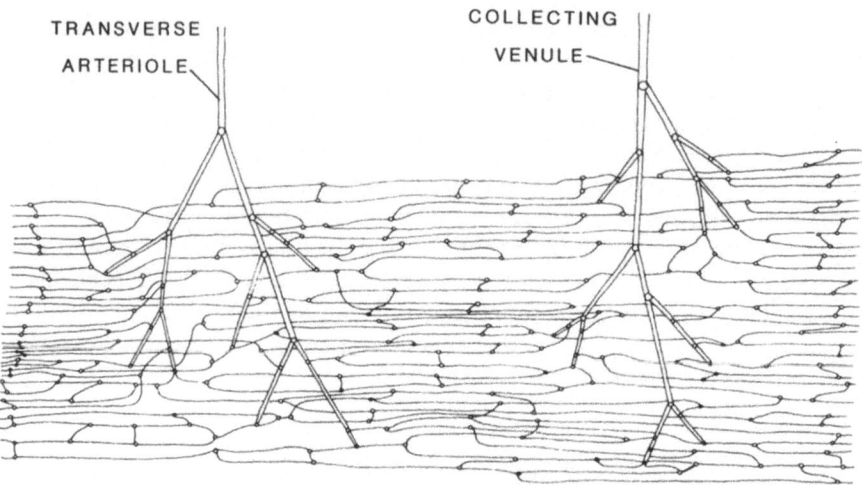

FIGURE 6.8. A schematic of the model capillary bundle element used in the network model. There are 30 parallel capillaries and 260 nodes. In this model the approriate number of capillary cross-connections (according to Eq. 6.10) are inserted at random, and the transverse arteriole and collecting venule are attached according to the model described in "The Transverse Arterioles and Their Venular Counterparts." This model forms the building block for the capillary bundle network.

connected to one another to terminate the capillary bundle. Due to the repeated arrangement of the capillary bundle, only a single capillary bundle element needs to be stored in the computer. For this purpose, all nodes are numbered, and vessel segment lengths, resting diameters, blood viscosity, and vessel distensibility are specified, as discussed in "Blood Flow in Single Vessels," above. Thus a capillary bundle is fed and drained by the capillary endings of n transverse arterioles and n collecting venules. In the current model n was chosen to be 6, so that the capillary bundle contains six arterioles, six venules, and a total of 1560 nodes. The next step is to assemble a number of capillary bundles together to form an entire muscle, and to connect these capillary bundles, by way of their transverse arterioles and collecting venules, to the network of arcade arterioles and venules. We start with the latter vessels.

In order to assemble the arcade network, the tissue volume must be specified. This is determined from the number of capillary bundles used, their average length, the average fiber/capillary ratio, which is close to 1 in the spinotrapezius muscle of mature rats, and the average muscle fiber and connective tissue cross-sectional area. The total number of bundle elements in a muscle will then determine its volume. Since the connective tissue in skeletal muscle makes up a small but finite fraction of the muscle, about 5–10% in rat skeletal muscle, it is important that such estimates be carried out on intact muscle specimens in which the collagen bundles between muscle fibers have

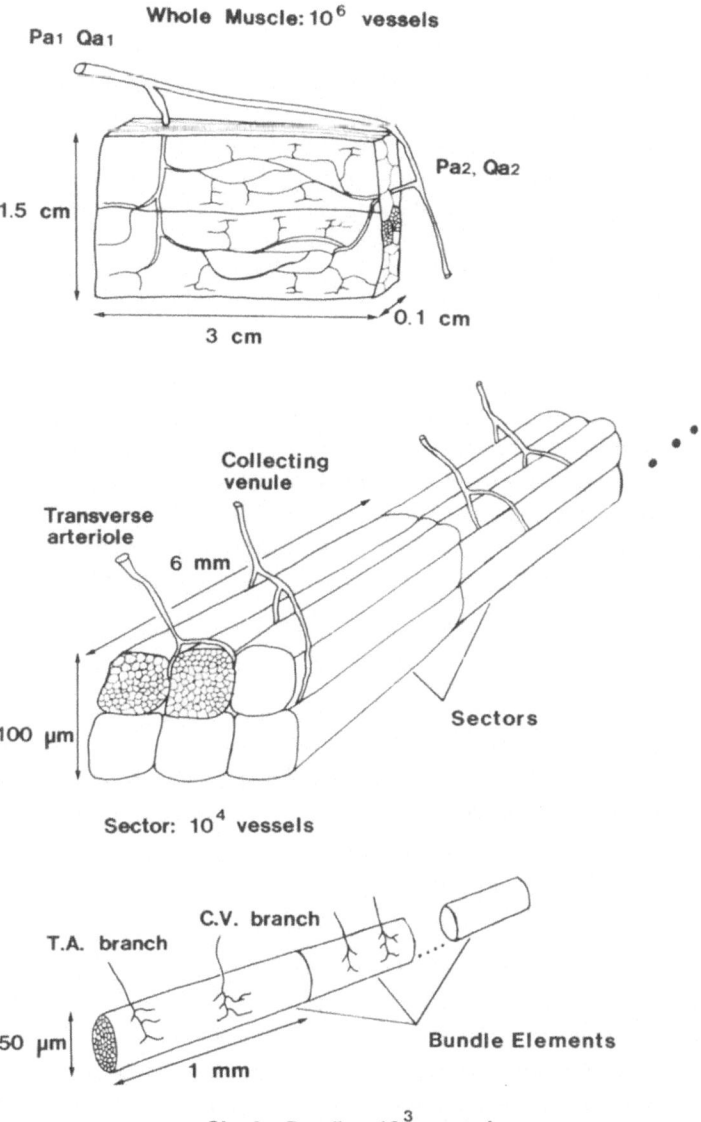

FIGURE 6.9. A schematic of the gracilis muscle model (*upper left*), a sector of the muscle model consisting of several muscle bundles (*upper right*), and a single capillary bundle (*bottom center*). The particular sector in the current model consists of six capillary bundles and is supplied by three transverse arterioles and drained by six collecting venules. Each capillary bundle consists of six capillary bundle elements and is connected to six transverse arteriolar branches and six collecting venule branches. In the numerical computations both the single bundles and the sectors are characterized by indefinite admittances, allowing pressure computations to be carried out for the entire muscle with about 1.2×10^6 microvessels.

not been disrupted by removal of the fascia or other surgical procedures (Mazzoni, 1985). Once the model tissue volume is specified, the arcade vessels are constructed using the arcade parameters listed in Table 6.1 (data are given in Engelson et al., 1985a, 1986). Since the number of these vessels in the rat spinotrapezius muscle is relatively small compared with the number of capillaries, the arcade network information is stored explicitly in the computer. The transverse arterioles and collecting venules are then connected at regular intervals, z (Table 6.2), into the arcade vessels. The venular network is constructed in an analogous way.

Network Equations

The flow in a single vessels is approximated, as discussed in "Blood Flow in Single Vessels," by Poiseuille's formula as given by Eq. 6.17. The viscosity is assumed to be a function only of shear rate in the vessel, according to Eq. 6.20 and the in vivo measurements by Lipowsky (1975). The vessel is assumed to be elastic with long-term steady behavior according to Eq. 6.18. To set up the network equations consider the flow at a single bifurcation. Conservation of mass requires that the inflow equals the outflow from the bifurcation. Thus for node i, we have

$$\sum_j \dot{Q}_{ij} = 0 \qquad (6.22)$$

where j is summed over the three neighboring nodes to node i. Flow into node i is positive and flow out of node i is negative. Integrating Eq. 6.17 for a fixed initial diameter d_{ij} over the length l_{ij} of a vessel and substituting Eqs. 6.17 and 6.20 into Eq. 6.22, we obtain:

$$\sum_j (p_i - p_j)G_{ij} = 0 \qquad (6.23)$$

where

$$G_{ij} = \frac{\pi d_{ij}{}^4}{8l_{ij}\left[k_1 + k_2\left(\dfrac{V_m}{d}\right)_{ij}^{-1/2}\right]^2}$$

and p_i is the pressure at node i, d_{ij} is the diameter of the vessel segment between node i and j, l_{ij} is the vessel segment length between these two nodes, and $(V_m/d)_{ij}$ is the shear rate in the vessel.

This system of equations for the nodal pressures is of order equal to the number of nodes. The system can be written in the form

$$GX = B,$$

where G is the conductance matrix, X is the vector of internal unknown nodal pressures, and B contains the specific flows at the feeding arterioles and draining venules as determined from the boundary pressures. Since only three vessels join at a node, there are only four nonzero elements in each row of G, leading to a sparse matrix if the total number of nodes is large. The system can be solved using any of several algorithms for sparse matrices, or simpler Gaussian elimination or decomposition methods. We implemented a sparse matrix algorithm proposed by Eisenstat et al. (1982) and updated by Vlach and Singhal (1983). In this routine, the number of operations and storage required to solve a $n \times n$ system grows linearly with n, as opposed to $n^3/3$, and n^2, for dense matrix inversion routines. An interative technique is used to account for the dependence of blood viscosity on shear rate and of vessel diameter on vessel pressure.

To account for the vessel distensibility and the non-Newtonian properties of the blood, the system is first solved for the internal nodal pressures at a given B corresponding to specific inflow and outflow pressures. The vessel diameters are then updated as a function of the newly computed vessel pressures, and the viscosities are updated according to the new shear rate in the vessels. A new conductance matrix G is formed, and the system is solved again for X. This procedure is repeated until all diameters, viscosities, and indefinite admittances converge to within 0.1% on successive iterations. Convergence usually required 15–18 iterations. The computation yields many details—for example, blood pressures at all nodes, X, flow in all vessels of the network, and the whole muscle pressure-flow relationship after repeated computations at different boundary pressures.

Numerical Procedures in Large Networks

In the numerical computations it is important to introduce a method to reduce the number of mathematical operations for the vast capillary network. This is readily achieved at the level of the transverse arterioles and collecting venules by means of *indefinite admittances* (Skalak, 1984). In this method each order 1 transverse arteriole and collecting venule is viewed as a "terminal" of the capillary bundle network. In the current case the capillary bundle is thus a 12-terminal network, and a 12×12 matrix of indefinite admittances can be defined so that each admittance, Y_{ij}, represents the flow into terminal i for a unit pressure applied at terminal j, while all other terminals are maintained at zero pressure. These admittances are obtained by direct computation for the capillary bundle, and can then be used to characterize the pressure-flow behavior of the bundle. The use of indefinite admittances allows a large number of bundles to be assembled together and connected to the arcade networks, without generating an intractably large system of equations for the whole muscle model.

We shall consider here two specific muscle models. The first is a model of a portion of rat spinotrapezius muscle, which permits comparisons of model predictions to in vivo measurements of pressure and flow within the network. This model, assembled according to the above procedure, consists of 15 capillary bundles, yielding a network with a total of 23,625 vessels of all hierarchies. The second model is larger and represents the entire rat gracilis muscle; it serves to compare the theoretical predictions with whole muscle pressure-flow data, but requires another level of anatomical complexity for its completion. In this thicker muscle, transverse arteriole and collecting venule trees are larger and each one supplies a larger number of capillary bundles. We continue to use the indefinite admittances, but apply them twice, once to a capillary bundle, as described above, and then again to a larger *sector* of microcirculation consisting of six capillary bundles, which is supplied by three transverse arterioles and drained by six collecting venules. Thus the transverse arterioles and collecting venules shown in Figure 6.8 for the individual capillary bundle are viewed in the gracilis model as branches of larger transverse vessels. The resulting gracilis muscle model is shown schematically in Figure 6.9. The model consists of 120 sectors, or a total of 720 capillary bundles, together with the arcade vessel network. The number of vessels in the whole muscle is thus about 1.2×10^6, but by dividing the muscle into repetitive n-terminal sectors and capillary bundles it is possible to achieve shorter computer times. Both the spinotrapezius and the gracilis muscle models have three to four feeding arterioles and draining venules.

Network Model Results

One of the important tests for a theory of blood flow is its ability to predict in vivo whole organ pressure-flow curves. More than any other experiment, it puts to a test our understanding of the microanatomy and vessel rheology and our ability to bridge the micro- and macrocirculation. Sutton (1987) has performed a series of experiments on the vasculature of the rat gracilis muscle using as perfusate both a Newtonian plasma substitute and suspensions of red blood cells of varying hematocrit. Figure 6.10 shows a comparison of model predictions for the gracilis muscle network model and these in vivo results. Also shown for comparison are data from Pappenheimer and Maes (1942) for perfusion of an isolated dog hind limb and Braakman (1988) for skeletal muscle. The pressure-flow relationship with a Newtonian fluid and dilated arterioles can be closely fitted by a third-order polynomial as predicted by Eq. 6.19. One of the characteristics of the pressure-flow relation in skeletal muscle is that the nonlinearity is not strong at most physiological flow rates, and is apparent only at low flow rates. In comparison, the pulmonary circulation shows stronger nonlinearities (Fung, 1984). This is due to the relatively small distensibility of skeletal muscle microvessels compared to the large distensibility of lung microvessels. The comparison between the results of the current theoretical model and the independent experiments for plasma is good (Fig. 6.10), considering the fact that the flow is normalized by the muscle mass. The comparison between theory and experimental results is also good at about 15% hematocrit, which includes the shear dependency of blood viscosity. The higher hematocrit causes a rightward shift of the pressure-flow curve and the shape of the curve at low flow rates becomes more nonlinear, as seen in the case of a single vessel (Fig. 6.7). Figure 6.11 shows the pressure-flow curve in the rat gracilis muscle for reconstituted blood at different feed

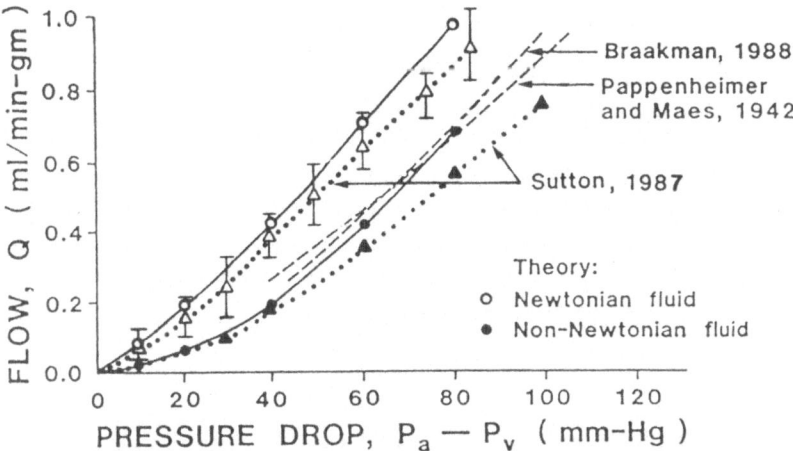

FIGURE 6.10. Comparison of rat skeletal muscle pressure-flow curves between experiment (*dashed lines*) and theory (*solid lines*). Agreement exists between the experiments by Sutton (1987) with plasma (*open triangles*) and at a hematocrit of about 15% (*closed triangles*) and the model predictions for Newtonian (*open circles*) and non-Newtonian (*closed circles*) blood. The bars on the experimental curve with plasma (*open triangles*) represent the range observed in 10 different animals, and reflect the anatomical variation among animals.

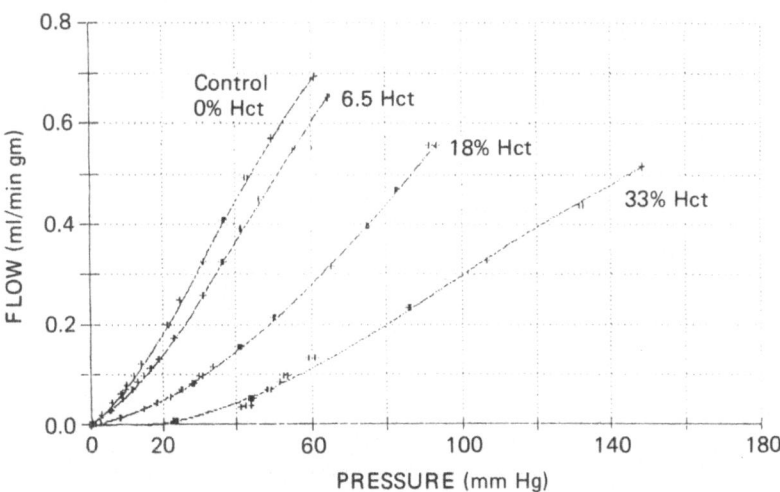

FIGURE 6.11. Experimental pressure-flow curves in rat gracilis muscle perfused with an autologous red cell suspension at different hematocrits. Increasing the hematocrit raises the organ flow resistance, resulting in a rightward shift of these curves.

hematocrits. Prediction of this curve requires independent analysis of cell distribution in the microvascular network, which is considered elsewhere (Schmid-Schönbein et al., 1980).

One of the attractive features of these network computations is that a complete picture of the hemodynamic quantities at any feed pressure can be obtained. Figure 6.12 shows a comparison between direct micropressure measurements in different hierarchies of microvessels in the rat spinotrapezius muscle (Zweifach et al., 1981) and the independent network model predictions. For this comparison the pressure is normalized with respect to the overall arterio-venous pressure drop so that

$$\bar{P} = \frac{P_{\text{vess}} - P_{\text{ven}}}{P_{\text{art}} - P_{\text{ven}}}$$

Thus the normalized pressure, $\bar{P}$, at the feeder arterioles is 1.0. The in vivo pressures were measured in mature Wistar-Kyoto rats under chloralose-urethane anesthesia (Zweifach et al., 1981). Model results are for an arterio-venous pressure drop, ΔP, of 38 mm Hg. Each point representing the model

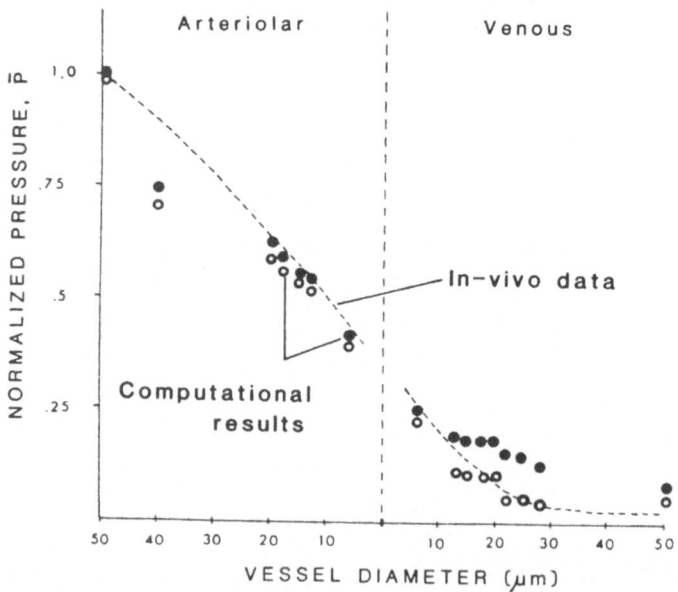

FIGURE 6.12. Micropressure distribution in arterioles, capillaries, and venules up to 50 μm in diameter in the rat spinotrapezius muscle. The pressure, $\bar{P}$, is normalized as described in the text. In vivo data are from Zweifach et al. (1981) for exteriorized spinotrapezius muscle. Computational results are shown for a Newtonian fluid (1.2 cp, *open circles*) and blood (*solid circles*). The shape of the curve indicates that the greatest pressure drop is located in the small arterioles. The low venular pressure drops are due mainly to the larger venular diameters and the high density of the venular arcades.

predictions represents the mean value among at least 10 vessels selected at random within each hierarchy, a procedure that closely simulates the selection of vessels during in vivo micropressure measurements. The theoretical pressure predictions for transverse arterioles, capillary, and small venules agree well with the in vivo data (Fig. 6.12). The discrepancy between the predicted and measured values at the level of the 40-μm arterioles is due to the fact that the normalized model boundary pressure of 1.0 drops greatly across the three single feeder arterioles to the position of the 40-μm arcade arterioles, which are typically located inside the muscle. An in vivo pressure of about 38 mm Hg is more typically found in a 50-μm arcade arteriole of the spinotrapezius muscle, so the pressure drop to the 40-μm vessels in vivo is not so pronounced.

Another test of our organ perfusion analysis is the capillary flow rate distribution. A realistic model should predict the distribution of flow accurately, since it is a primary determinant of muscle performance and tissue viability. Figure 6.13 shows the frequency distribution of capillary flow for 100 capillaries selected at random throughout the network model for spinotrapezius muscle, for an arteriovenous pressure drop of $\Delta P = 38$ mm Hg with about 15% hematocrit. The magnitudes of the capillary flows are in the same range as those observed in vivo (Zweifach et. al., 1981). Capillary flows in different regions of the same muscle show a relatively narrow dispersion, considering the large volume of these muscles and the widely differing dis-

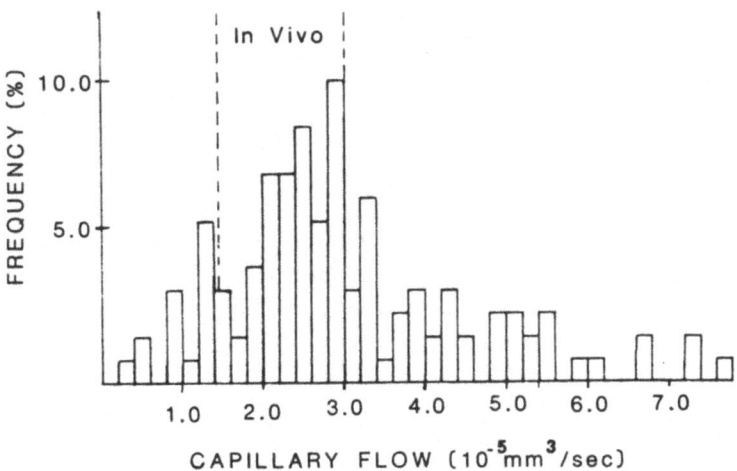

FIGURE 6.13. Frequency distribution of capillary flow rate for 100 randomly selected capillaries in the spinotrapezius muscle model with blood perfusate at normal perfusion pressures. Magnitudes of the computed flows are in the same range as the in vivo flows, although the model predictions have a tendency to show larger dispersion than the in vivo data. The dispersion of the capillary flows as predicted by the model is sensitive to capillary length, diameter, and position of the capillaries in the network.

tances of capillaries to the feeder arterioles. The network model furthermore allows us to investigate how sensitive the flow dispersion is with respect to capillary length, diameter, and vessel position in the capillary bundles. It is interesting to note that besides diameter and length, the cross-connections between the parallel capillaries within the capillary bundles (see Table 6.3) play an important role in capillary flow rate dispersion. The fewer cross-connections there are, the smaller the dispersion, and vice versa. Such cross-connections may often have flow rates that are substantially less than the mean capillary flow rates, which leads to wider dispersions in the histograms. In vivo estimates of flow are preferentially made in straight sections of the parallel capillaries, so that model flow predictions less than the in vivo estimates may represent long capillaries or cross connections between vessels of similar pressure. Flow predictions higher than the in vivo values may represent thoroughfare channels that receive higher than average flow due to slightly larger diameters and higher pressure drops than the average capillaries.

Hemodynamic Properties of the Capillary Network

In this section we will illustrate, with some examples, how the network model may serve to examine the hemodynamic consequences of specific regulatory adjustments. As an example we will demonstrate the effects of variations in the transverse arteriolar diameters on the spatial distribution of blood pressure and flow in a capillary bundle. Transverse arterioles usually have the highest smooth muscle tone in the microcirculation (Schmid-Schönbein et al., 1987b), and during normal autoregulatory control or vasomotion they may completely occlude their lumen. Our results can demonstrate the operation of a hemodynamic capillary unit, such as the one recently suggested by Delashaw and Duling (1988) based on in vivo observations. Consider a capillary bundle of seven capillary elements, with each element as shown in Figure 6.8. We have examined the effect of a diameter reduction in one or more transverse arteriolar vessels while maintaining a constant pressure at all feeding and draining vessels to the bundle at 40 mm Hg and 10 mm Hg, respectively. Flows and pressures were first computed with all transverse arterioles having the same resting diameter, and subsequently compared with the situation when one or more transverse arterioles had reduced diameters.

   Figure 6.14A and 6.14B show the capillary pressure and flow distributions in bundle element 4 located at the center of the seven-element capillary bundle with open arterioles. The flow rates are within the range of in vivo measurements and are uniform among the different bundle elements. Figure 6.14C and 6.14D show the pressure and flow distributions in the same element after an 80% decrease in the resting diameter of the transverse arteriolar in element 4. The mean capillary flow in this bundle element is 70% lower than before the arteriolar narrowing, and the change in mean capillary pressure is only about 35% in this extreme constriction. This is remarkable and serves to balance

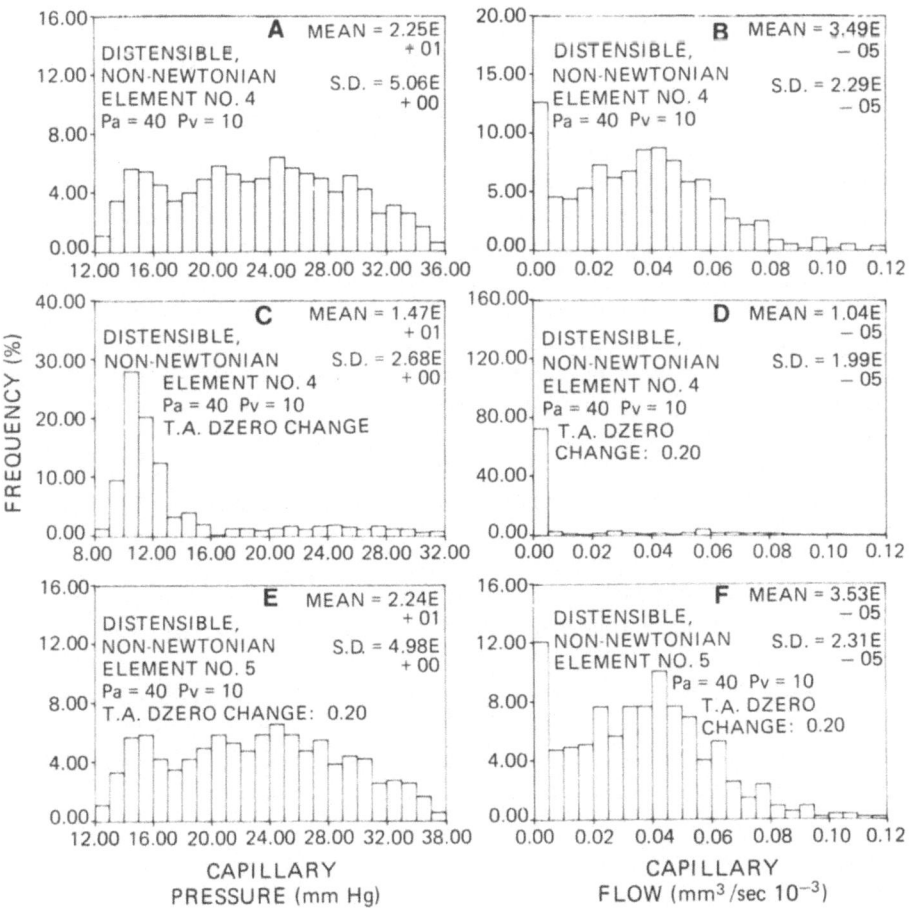

FIGURE 6.14. Capillary pressure histograms (*left column*) and capillary flow histograms (*right column*) in a single capillary bundle consisting of seven bundle elements. In panels *A* and *B* histograms are shown for element 4 (the center element of the bundle) at normal transverse arteriolar diameters. In panels *C* and *D*, all arteriolar input and venular output pressures were held constant, but the diameter of the transverse arteriole in element 4 was reduced by 80%, as during vasomotion. Whereas the flows fall to almost zero values, the pressure is maintained above the pressures in the neighboring collecting venules, thereby preserving finite filtration pressures in the capillaries during the constriction phase of the vasomotor cycles. In panels *E* and *F*, the pressure and flow in the neighboring bundle element 5 are shown at the time of arteriolar narrowing to element 4. Its pressure and flow distribution are virtually identical to the control situation (panels *A* and *B*) before constriction, which demonstrates the localized influence of the transverse arterioles to their own capillary network.

Starling filtration pressures during vasomotor activity of the transverse arterioles. It is only possible by the close capillary connections in the capillary bundles.

An interesting consequence of the arrangement of repeated capillary bundle elements is that each transverse arteriole can control the flow in its local tissue region, without affecting the flow in the neighboring tissue. Figure 6.14E and 6.14F show the pressure and flow distribution in bundle element 5, after arteriolar narrowing in element 4. In this neighboring element the mean flow has decreased by only 1%, and the distribution is nearly the same as in element 4 before the arteriolar narrowing (Fig. 6.14B). The mean pressure decreased only by 0.4% in bundle element 5. If two or more transverse arterioles are constricted, the changes in capillary pressure and flow are also restricted to the capillary bundle elements that they supply.

Microvascular Flow During Unsteady Pressures

Direct measurements of the pressure-flow relationship in skeletal muscle of different species (Sutton, 1987; Braakman, 1988) suggest that in spite of negligibly small inertial forces, time-dependent effects are present. In rat skeletal muscle, without vasomotor control, there appear to be two causes for these unsteady effects, the viscoelastic properties of the blood vessels and the dynamic intraction between the distensible wall and the viscous fluid in the presence of unsteady arterial pressures. The latter case has been designated as *pulsatile viscous flow* (Schmid-Schönbein et al, 1989). We shall discuss here in more detail the specific case of an elastic microvessel.

Consider the short-term elastic response of a microvessel such that Eq. 6.12 reduces to the form:

$$P = \alpha E \tag{6.24}$$

where $\alpha = \alpha_1 + \alpha_2$. Let A be the cross-sectional area; then

$$E = \frac{1}{2}\left(\frac{A}{A_0} - 1\right) \tag{6.25}$$

where A_0 is the reference cross-section. Conservation of mass for a distensible, impermeable vessel segment with incompressible fluid is

$$\frac{\partial \dot{Q}}{\partial z} = -\frac{\partial A}{\partial t} \tag{6.26}$$

This equation leads to viscous propagation in the microcirculation so that $\dot{Q}$, P, E, and A are functions both of time and axial position z. Substitution of Eqs. 6.24 and 6.25 into Eq. 6.26 gives

$$\frac{\partial \dot{Q}}{\partial z} = -\frac{2A_0}{\alpha}\frac{\partial P}{\partial t} \tag{6.27}$$

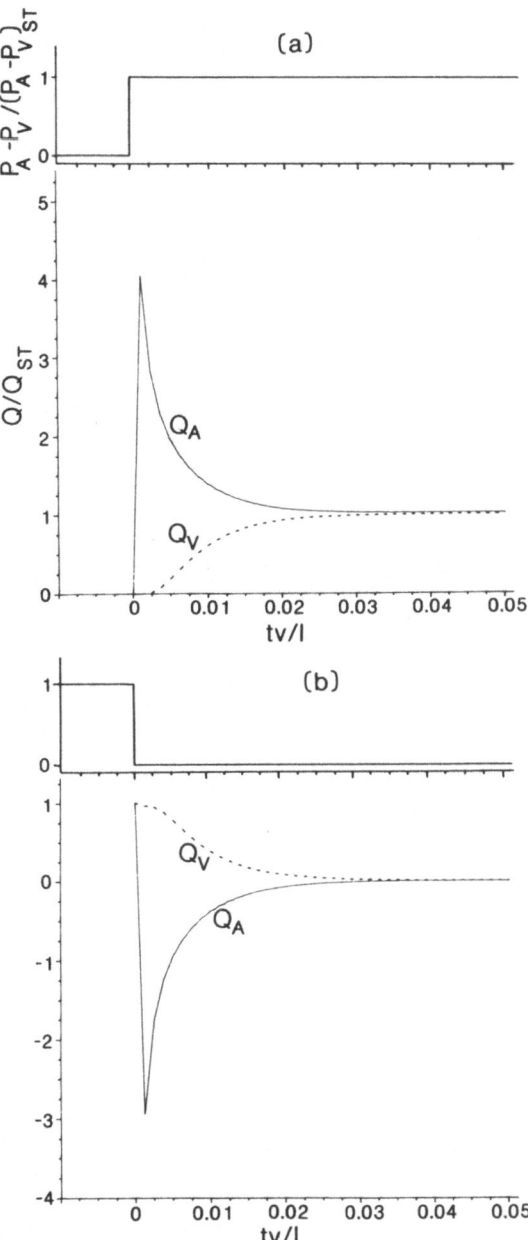

FIGURE 6.15. Upstream flow, Q_A, and downstream flow response, Q_V, to a step arterial pressure increase (*top*) and decrease (*bottom*) in a single microvessel. The flow is normalized by the steady state flow, Q_{ST}, and the time t by the mean velocity, V, at steady state and by the vessel length l. For such pulsatile viscous flow the upstream and downstream flow are not always equal, due to local vessel expansion. During such step pressures a prominent flow overshoot is observed in agreement with experimental observations (Braakmann, 1988; Schmid-Schönbein et al., 1989). (Reprinted with permission from Biorheology, 26, Schmid-Schönbein et al., Dynamic viscous flow in distensible vessels of skeletal muscle microcirculation: application to pressure and flow transients, 1989, Pergamon Press plc.)

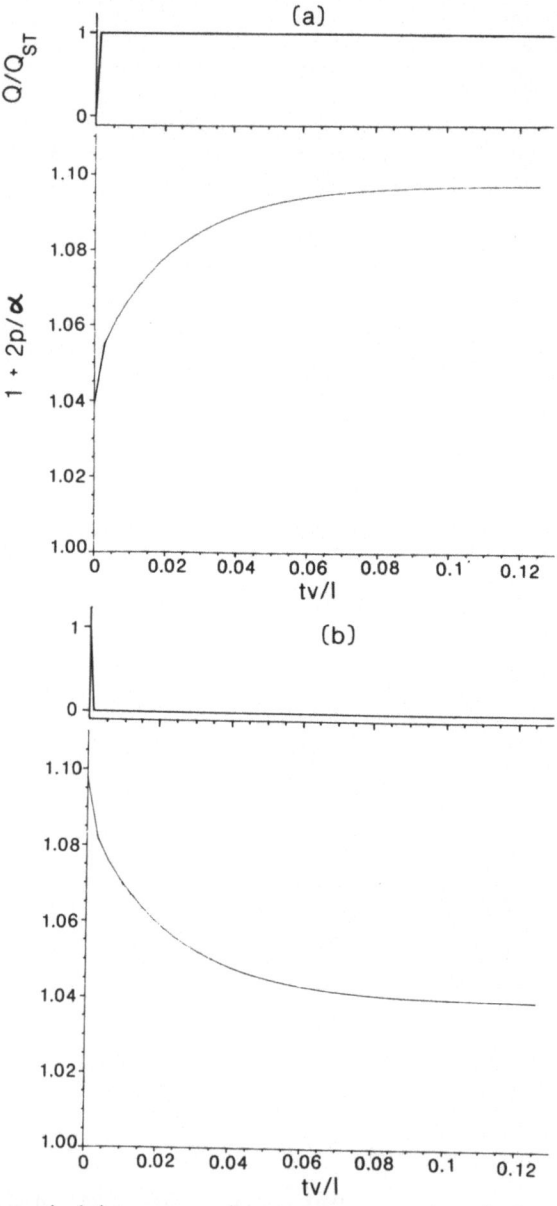

FIGURE 6.16. Theoretical time course of upstream pressure in a single microvessel after a step flow increase (*top*) and decrease (*bottom*). Q_{ST} is the steady state flow rate, the pressure p is presented in a normalized form as $1 + 2p/\alpha$, where α is the vessel stiffness as defined in Eq. 6.24. The time along the abscissa is also normalized by the mean velocity V and by the vessel length l. The gradual rise in upstream pressure is due to the delayed development of the full flow along the length of the vessel, after the initial vessel expansion along its upstream segment. Following upstream flow reduction (bottom) the vessel continues to recoil, giving rise to a continued outflow and viscous pressure drop along its length. (Reprinted by permission from Biorheology, 26, Schmid-Schönbein et al., Dynamic viscous flow in distensible vessels of skeletal muscle microcirculation: application to pressure and flow transients, 1989, Pergamon Press plc.)

so that after differentiation of Eq. 6.17 with respect to z we find the governing equation in the form of

$$\frac{\partial P'}{\partial t} = C^2 \frac{\partial^2 P'^3}{\partial z^2} \tag{6.28}$$

where $C^2 = A_0 \alpha / 48 \pi \mu$ and $P' = 1 + 2P/\alpha$ is the normalized pressure. Eq. 6.28 can also be rewritten in the form of a diffusion equation for P'^3 with nonlinear coefficients:

$$\frac{\partial P'^3}{\partial t} = (3C^2 P'^2) \frac{\partial^2 P'^3}{\partial z^2} \tag{6.29}$$

The coefficient C^2 is of the order of 0.34 to 95.2 cm^2/sec for arterioles and capillaries, respectively, indicating that the unsteady term on the left-hand side is comparable with the term on the right-hand side of Eq. 6.29. In contrast to wave propagation in large arteries, no inertial forces are present in these small vessels.

A solution to Eq. 6.29 can be obtained readily with finite difference approximations. Figures 6.15 and 6.16 show the results for a step arterial

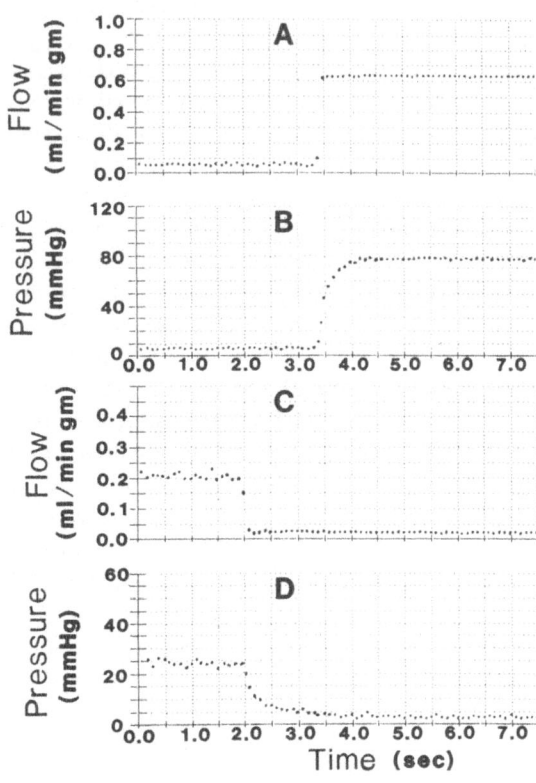

FIGURE 6.17. Experimental pressure and flow histories for a step flow increase (panels A and B) and decrease (panels C and D) in the rat gracilis muscle according to Sutton (1987). Note the delayed pressure histories, in agreement with the predictions in Figure 6.16.

pressure and a step arterial flow, respectively. In the presence of a step pressure the arterial flow shows an overshoot due to expansion of the upstream vessel segment. The venous flow develops only gradually without overshoot during the increase of the transient to its steady state value. A step flow in turn results in a gradual increase of the arterial pressure due to a gradual buildup of the viscous pressure gradient along the tube. Figure 6.17 shows experimental results from the gracilis muscle using plasma (Sutton, 1987) that qualitatively agree with the predictions by the pulsatile viscous flow theory. In general, for rat skeletal muscle with plasma, these transients are limited to no more than a few seconds. With whole blood or in larger muscles of larger species, longer transients are expected.

The Zero-Flow Arterial Pressure

In our analysis carried out to this point, we have seen that all steady state pressure-flow curves with plasma or with dispersed red cell suspensions pass through the origin. However, in the gracilis muscle perfusion experiments the arterial pressure-flow curves do not pass through the origin under all circumstances (Sutton and Schmid-Schönbein, 1989). A nonzero pressure intercept has been reported repeatedly and was denoted as *zero-flow pressure* (ZFP). Burton (1951) called it "critical closing pressure." The literature on this subject is large. We have recently reviewed the previous hypotheses on the mechanisms of the phenomenon and subjected the gracilis muscle preparation to a thorough analysis of the ZFP (for references see Sutton and Schmid-Schönbein, 1989). In the following we will summarize our findings.

The shape of the pressure-flow curve is sensitive to the rate at which zero flow is reached. For this reason it is convenient to distinguish between a steady state and a transient pressure reduction to the ZFP. It is also important to note for how long the tissue is kept at zero pressure.

Under steady state conditions (periods between about 30 s and about 2 min) the rat gracilis muscle shows a ZFP between 3 and 15 mm Hg with red cell suspensions between 0 and 40% hematocrit. As long as the red cells are not aggregated the particular ZFP is largely independent of hematocrit in this range. It is, however, sensitive to the pressure in the *central* circulation. When the central blood pressure of the animal is reduced to zero, the ZFP in the gracilis muscle drops instantaneously to zero (Fig. 6.18). Detailed analysis shows that this steady state ZFP is due to an inflow from the central circulation to the gracilis muscle circulation via a hidden arterial arcade anastomosis to the underlying muscle. This was found in every experiment. If the gracilis muscle is perfused with a feedback controlled pump at zero arterial pressure, a negative arterial flow is observed during normal central blood pressure (Fig. 6.18). This clearly shows the presence of an inflow in spite of a careful vessel isolation procedure. The influence of the hidden arteriolar anastomosis can

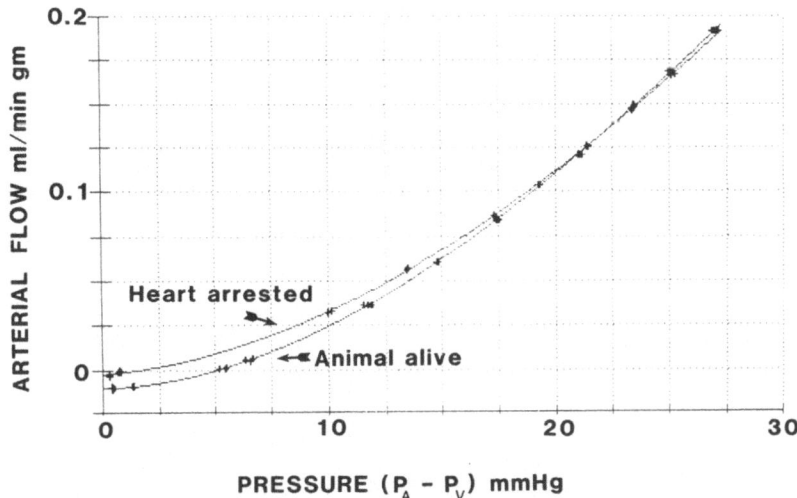

FIGURE 6.18. Arterial flow in the vasodilated resting rat gracilis muscle as a function of the arteriovenous pressure drop. Perfusion is maintained at steady state up to 2 min at each measurement point. The muscle is perfused with a reconstituted Newtonian viscous albumin solution according to Sutton (1987). The two curves are generated with the same muscle preparation within minutes before (*lower curve*) and after (*upper curve*) cardiac arrest. The lower curve shows a positive zero-flow pressure due to inflow from a hidden arterial anastomosis to the arcade arterioles in a neighboring muscle. It shows negative arterial flow (backflow) at zero arterial pressure due to the anastomosis. The upper curve passes through the origin within experimental error.

also be readily demonstrated with the network model, as shown in Figure 6.19. The network model has shown that the anastomosis must feed into the arteriolar arcades. Capillary or venular anastomoses do not raise the arterial ZFP by more than a fraction of a millimeter of mercury, due to the low venular resistance in the muscle microcirculation. The hidden anastomosis is not necessarily the only cause for a positive ZFP. Under steady state, the ZFP can also be elevated by aggregation of the red cells into rouleauxs with dextran. In the microcirculation rouleauxs may be able to store elastic energy before breakup into smaller rouleauxs or single cells.

All of the above perfusion conditions occur without sign of vessel collapse. Direct fixation of the muscle during perfusion and observation of histological sections shows that all vessels of the microcirculation have an open lumen. If zero pressure is maintained for prolonged periods, more than about 5 min, however, individual capillaries reduce their lumen cross-section and develop rounded endothelial nuclei with obstructed lumen. Arterioles and venules maintain an open lumen under these conditions.

Finally we turn to the unsteady perfusion. As we saw in the previous section,

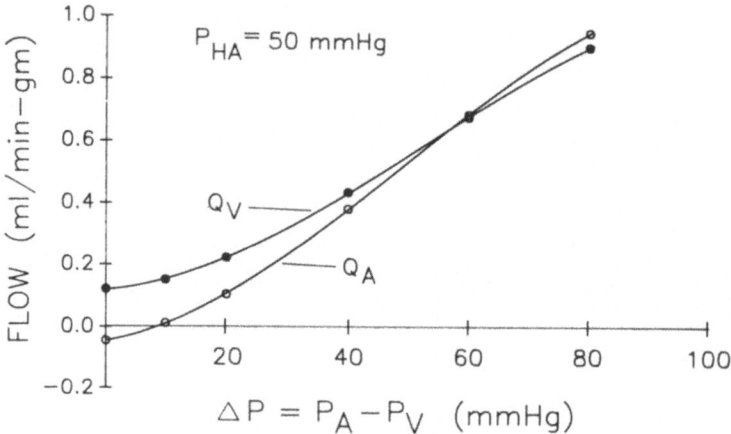

FIGURE 6.19. Theoretical pressure-flow relation for a model gracilis muscle vascular network with an additional "hidden" arterial anastomosis connected to the arcade arterioles. The pressure in the hidden anastomosis, p_{HA}, was kept at 50 mm Hg. Note that the arterial flow, Q_A, passes through the zero flow axis at a finite pressure, and gives negative arterial flows below the zero-flow pressure. The venous outflow, Q_V, is positive even for zero pressure gradient due to input from the hidden anastomosis.

when the arterial flow is instantaneously reduced to zero, the arterial pressure requires a prolonged period to reach steady state (Figs. 6.15 and 6.17). Thus under unsteady perfusion conditions, significantly higher but transient ZFPs can be seen depending on how high the initial pressure was and at what time the reading is made. Under a normal pulse rate such transients have usually not completely decayed so that steady state ZFPs are not observed. This may explain some of the extraordinary high critical closing pressures that have been observed in the past.

Conclusion

The analysis presented in this chapter shows the feasibility of predicting in vivo whole organ pressure-flow curves from microvascular network models. Basic information about the microanatomy, the microvessel distensibility, and blood rheology are necessary so that a realistic analysis can proceed and critical tests of the model predictions can be carried out.

Skeletal muscle in the dilated state has relatively stiff microvessels, resulting in a pressure-flow relationship that is almost linear in the normal flow range but nonlinear at low flows. In the presence of a Newtonian plasma fluid medium, the nonlinear pressure-flow curve is due to vessel distensibility only, whereas in the presence of blood cells the pressure-flow curve is shifted toward higher resistances. The capillary flow distribution is remarkably uniform throughout the muscle, in large part due to the arteriolar and venular arcades. Transverse arterioles have a strong but localized control over capillary flow, with a weaker influence on capillary pressures due to the particular arrangement of the capillary bundles. The zero-flow pressure at the arterial inflows in the rat gracilis muscle has multiple causes, including an arteriolar anastomosis with an underlying muscle, aggregation of blood cells, and transient changes in arterial blood flow. Future developments should include other factors that influence microvascular perfusion, such as autoregulatory controls, cell distributions, and skeletal muscle contraction.

Acknowledgment. This research was supported by National Science Foundation grant PCM-85-16662 and in part by U.S. Public Health Service grants HL-10881 and HL-39680.

References

Atabeck HB (1980) Blood flow and pulse propagation in arteries. In Patel DJ, Vaishnav RN (eds) *Basic Hemodynamics and its Role in Disease Processes.* University Park Press, Baltimore, pp 255–361.

Bauer RD, Busse R, Schabert A (1985) The input impedance of the peripheral vascular termination in skeletal muscle. *Pflugers Arch* 403:308–311.

Braakmann R (1988) Pressure-flow relationships in skeletal muscle. PhD Dissertation, University of Amsterdam.

Burton AC (1951) On the physical equilibrium of small blood vessels. *Am J Physiol* 164:319–329.

Chen IIH (1983) A mathematical representation for vessel network II. *J Theoret Biol* 104:647–654.

Chen IIH (1984) A mathematical representation for vessel notwork III. *J. Theoret Biol* 111:115–121

Delashaw JB, Duling BR (1988) A study of the functional elements regulating capillary perfusion in striated muscle. *Microvasc Res* 36:162–171.

Eisenstat SC, Gursky MC, Schultz MH, Sherman AH (1982) Yale sparse matrix package I: the symmetric codes. *Int J Num Meth Eng* 18:1145–1151.

Engelson ET, Schmid-Schönbein GW, Zweifach BW (1985a) The microvasculature in skeletal muscle. III. Venous network anatomy in normotensive and spontaneously hypertensive rats. *Int J Microirc Clin Exp* 4:229–248.

Engelson ET, Schmid-Schönbein GW, Zweifach BW (1986) The microvasculature in skeletal muscle. II. Arteriolar network anatomy in normotensive and hypertensive rats. *Microvasc Res* 31:356–374.

Engelson ET, Skalak TC, Schmid-Schönbein GW (1985b) The microvasculature in skeletal muscle. I. Arteriolar network in rat spirotrapezivs muscle. *Microvasc Res* 30:29–44.

Fung YC (1978) Mechanical properties of blood vessels. In PC Johnson (ed) *Peripheral Circulation.* John Wiley & Sons, New York, pp 45–79.

Fung YC (1984) *Biodynamics: Circulation.* Springer Verlag, New York.

Granger H, Meininger GA, Borders JL, Morff RJ, Goodman AH (1984) Microcirculation of skeletal muscle. *Phys Pharm Microcirc* 2:181–265.

Hudlická O (1973) *Muscle Blood Flow. Its Relation to Muscle Metabolism and Function.* Swets and Zeitlinger BV, Amsterdam.

Koller A, Dawant B, Liu A, Popel AS, Johnson PC (1987) Quantitative analysis of arteriole network architecture in cat sartorius muscle. *Am J Physiol* 253:H154–H164.

Krogh A (1919) Number and distribution of capillaries in muscle with calculation of oxygen pressure head necessary for supplying the tissue. *Am J Physiol* 52:409–415.

Lee J, Salathè EP, Schmid-Schönbein GW (1987) Fluid exchange in skeletal muscle with viscoelastic blood vessels. *Am J Physiol* 253:H1548–H1566.

Lindbom L, Arfors KE (1985) Mechanism and site of control for variation in the number of perfused capillaries in skeletal muscle. *Int J Microcirc Clin Exp* 4:19–38.

Lipowsky HH (1975) In-vivo studies of the rheology of blood in the microcirculation. PhD Dissertation, University of California, San Diego. University Microfilms No. 75-29446, Ann Arbor, MI.

Mazzoni MC (1985) The effect of skeletal muscle deformation on lymphatic vessels. Master of Science Thesis, University of California, San Diego.

Pappenheimer JR Maes JP (1942) A quantitative measure of the vasomotor tone in the hind limb muscle of the dog. *Am J Physiol* 137:187–199.

Popel AS (1987) Network models of peripheral circulation. In Skalak R, Chien S (Eds) *Handbook of Bioengineering.* McGraw-Hill, New York, pp 20.1–20.24.

Popel AS, Torres-Filho IP, Johnson PC, Bouskela E (1988) A new scheme for hierarchical classification of anastomosing vessels. *Int J Microcirc Clin Exp* 7:131–138.

Schmid-Schönbein GW (1988) A theory of blood flow in skeletal muscle. *J Biomech Eng* 110:20–26.

Schmid-Schönbein GW, Firestone G, Zweifach BW (1986b) Network anatomy of arteries feeding the spinotrapezius muscle in normotensive and hypertensive rats. *Blood Vessels* 23:34–49.

Schmid-Schönbein GW, Lee SY, Sutton D (1989) Dynamic viscous flow in distensible vessels of skeletal muscle microcirculation: application to pressure and flow transients. *Biorheology.* 26:215–227.

Schmid-Schönbein GW, Murakami H (1985) Blood flow in contracting arterioles. *Int J Microcirc Clin Exp* 4:311–328.

Schmid-Schönbein GW, Skalak TC, Engelson ET, Zweifach BW (1986a) Microvascular Network Anatomy in Rat Skeletal Muscle. In Popel AS, Johnson PC (eds) *Microvascular Network: Experimental and Theoretical Studies.* Karger, Basel, pp 38–51.

Schmid-Schönbein GW, Skalak TC, Firestone G (1987a) The microvasculature in skeletal muscle. V. The arteriolar and venular arcades in normotensive and hypertensive rats. *Microvasc Res* 34:385–393.

Schmid-Schönbein GW, Skalak R, Usami S, Chien S (1980) Cell distribution in capillary networks. *Microvasc Res* 19:18–44.

Schmid-Schönbein GW, Zweifach BW, DeLano FA, Chen P (1987b) Microvascular tone in a skeletal muscle of spontaneously hypertensive rats. *Hypertension* 9:H1548–H1566.

Skalak TC (1984) A mathematical hemodynamic network model of the microcirculation in skeletal muscle, using measured blood vessel distensibility and topology.

PhD Dissertation, University of California, San Diego. University Microfilms DA-8418303, Ann Arbor, MI.

Skalak R, Özkaya N, Secomb TW (1986) Biomechanics of capillary blood flow. In Schmid-Schönbein GW, Woo SCY, Zweifach BW (eds) *Frontiers in Biomechanics.* Springer-Verlag, New York, pp 299–313.

Skalak TC, Schmid-Schönbein GW (1986a) The microvasculature in skeletal muscle. IV. A model of the capillary network. *Microvasc Res* 32:333–347.

Skalak TC, Schmid-Schönbein GW (1986b) Viscoelastic properties of microvessels in rat spinotrapezius muscle. *J Biomech Eng* 108:193–200.

Spalteholz W (1888) Die Vertheilung der Blutgefässe im Muskel. *Abh Sächs Ges Wiss Math Phys* 14:509–528.

Sutton DW (1987) The pressure-flow relationship in the isolated gracilis muscle of the rat. PhD Dissertation, University of California, San Diego.

Sutton DW, Schmid-Schönbein GW (1989) Hemodynamics at low flow in the resting, vasodilated rat skeletal muscle. *Am J Physiol.*, in press.

Vlach J, Singhal K (1983) *Computer Methods for Circuit Analysis and Design.* Van Nostrand Reinhold Company, New York.

Zweifach BW, Kovalcheck S, Delano FA, Chen P (1981) Micropressure-flow relationships in a skeletal muscle of spontaneously hypertensive rats. *Hypertension* 3:601–614.

Part II Pulmonary Microvascular Mechanics

7
Pulmonary Capillary Blood Pressure

SCOTT A. BARMAN, DORIS COPE, RONALD C. ALLISON, and
AUBREY E. TAYLOR

Introduction

When one considers the various vascular pressures throughout the body's
circulation, it is remarkable that the lungs' circulation is such a high-flow
circuit with low resistance. In addition, in various types of lung pathology it
is even more remarkable that the alveoli do not easily fill with fluid, which
would compromise the gas exchange function of the lung. For many years,
it was thought that the pulmonary microvascular pressure was quite low
because the average pulmonary systolic pressure is about 20–25 mm Hg
and pulmonary diastolic pressure is approximately 10–15 mm Hg. Since left
atrial pressure is only 0–5 mm Hg, capillary pressure was estimated to be
approximately 5–10 mm Hg. However, it was not until 1967 that Gaar
et al. applied the isogravimetric technique developed by Pappenheimer and
Soto Rivera (1948) to the isolated dog lung to measure pulmonary capillary
pressure.

Pulmonary Capillary Pressure Measurements

Isogravimetric Capillary Pressure Determination

Gaar et al. (1967) used a linear electrical analog that was based on the
assumption that the pulmonary arterial pressure (P_{pa}) dropped across a precapillary or pulmonary arterial resistance (R_{pa}) to a capillary pressure (P_{pc}).
Subsequently, P_{pc} dropped to a venous pressure (P_{pv}, or left atrial pressure in
in vivo lungs) across a postcapillary or pulmonary venous resistance (R_{pv})
(shown in Fig. 7.1). Using Papenheimer and Soto Rivera's (1948) weighing
procedures, they calculated capillary pressure by the following technique: first,
the isolated lung became isogravimetric (i.e., not gaining or losing weight) by
altering P_{pa} and P_{pv}. At this condition, the pulmonary arterial and venous
pressures and blood flow ($\dot{Q}_B$) were measured. Figure 7.2A shows a measure

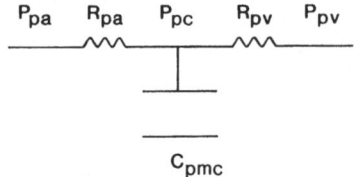

P_{pa} R_{pa} P_{pc} R_{pv} P_{pv}

C_{pmc}

FIGURE 7.1. Electrical analog of the pulmonary circulation. P_{pa}, pulmonary arterial pressure; P_{pc}, pulmonary capillary pressure; P_{pv}, pulmonary venous pressure; R_{pa}, precapillary resistance; R_{pv}, postcapillary resistance; C_{pmc}, pulmonary microvascular capacitance.

of isogravimetric capillary pressure. The pressure drops across R_{pa} and R_{pv} can be described mathematically as:

$$P_{pa} - P_{pc} = R_{pa} \times \dot{Q}_B \quad or \quad P_{pc} = P_{pa} - (R_{pa} \times \dot{Q}_B) \qquad (7.1)$$

$$P_{pc} - P_{pv} = R_{pv} \times \dot{Q}_B \quad or \quad P_{pc} = P_{pv} + (R_{pv} \times \dot{Q}_B) \qquad (7.2)$$

When $\dot{Q}_B$ is decreased, P_{pa} will fall and the lung will tend to lose weight because P_{pc} has decreased. To counteract this tendency to lose weight, venous pressure is elevated to increase P_{pc} until the lung is in a new isogravimetric state. This procedure was repeated seven times in the experiment depicted in Figure 7.2A, wherein first $\dot{Q}_B$ was decreased, followed by a decrease in P_{pa}, and then P_{pv} was subsequently elevated to maintain the lung in a constant weight condition. The upper curve represents Eq. 7.1 and the slope of the line is R_{pa} with the intercept on the pressure axis being equal to P_{pc} when $\dot{Q}_B = 0$. Also, the lower curve represents Eq. 7.2. and its slope is R_{pv}. As seen in the upper curve, when $\dot{Q}_B = 0$, the pressure intercept also is P_{pc}. Gaar et al. (1967) used this technique to measure pulmonary capillary pressure, but since it required weighing the lung and maintaining it in an isogravimetric state, each capillary pressure determination was difficult to accomplish since each isogravimetric point sometimes required 10–15 min to obtain a reliable isogravimetric condition. Therefore, this technique for measuring P_{pc} was used in only a few studies in the pulmonary circulation. However, the isogravimetric technique allowed Gaar et al. to derive the following equation, which described the relationship of pulmonary capillary pressure to pulmonary arterial pressure, pulmonary venous pressure, and the ratio of precapillary to postcapillary resistances in the lung:

$$P_{pc} = P_{pv} + (R_{pv}/R_{pa} + R_{pv})(P_{pa} - P_{pv}) \quad or$$

$$P_{pc} = P_{pv} + 0.4\,(P_{pa} - P_{pv}) \qquad (7.3)$$

Since R_{pa} and R_{pv} probably change in many physiological and pathological conditions, the obvious constraints that were present when measuring P_{pc} needed to be remedied before this important determinant of lung fluid balance could be determined.

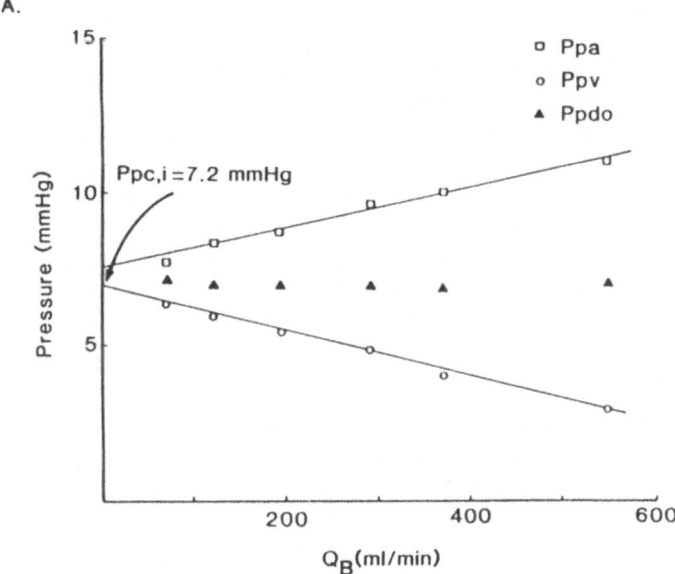

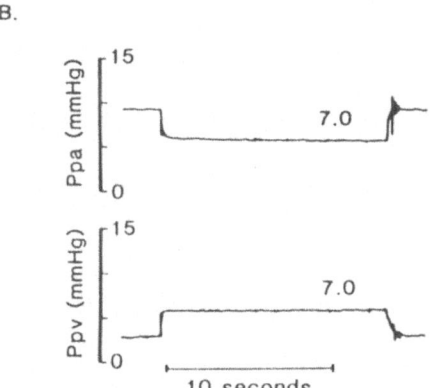

FIGURE 7.2. *A*: Isogravimetric technique for determining isogravimetric capillary pressure ($P_{pc,i}$). Isogravimetric arterial (*squares*) and venous (*circles*) pressures (P_{pa} and P_{pv}) are plotted as a function of isogravimetric blood flow ($\dot{Q}_B$) in the isolated blood-perfused lung. *B*: Double vascular occlusion technique for determining capillary pressure. Pressure-time tracings show that during double-occlusion arterial (P_{pa}) and venous (P_{pv}) perfusion pressures rapidly equilibrate to P_{pc}. (Modified from Townsley et al. 1986.)

Double-Occlusion Capillary Pressure Measurement

The remedy finally came about when the classic works of Hakim et al. (1979), Dawson et al. (1982), Parker et al. (1983), and Ryan et al. (1980) were published. These investigators showed that clamping of the arterial or venous pressures simultaneously or individually yielded very good estimates of pulmonary capillary pressures when the pressure transients following arterial inflow clamping or venous outflow clamping were properly analyzed. Figure 7.2B shows the quickest and easiest way to accurately assess P_{pc}. In an isolated lung when both P_{pa} and P_{pv} are simultaneously clamped, the pressures equilibrate to the same value (7.0 mm Hg in this case). Therefore, Figure 7.2 illustrates that the measurement of P_{pc} is the same whether the double-clamp technique or the classic Gaar isogravimetric technique is used. The double-clamp capillary pressures (closed triangles) were measured at each isogravimetric state by simultaneously clamping P_{pa} and P_{pv} as shown in Figure 7.2A. Note that the double-clamp pressures in every instance were very close to the isogravimetrically determined capillary pressure (pressure intercept of the P_{pa} and P_{pv} regression lines). Thus, the double-clamp technique provided a new means for the easy measurement of pulmonary capillary pressure and suggested that the pulmonary microcirculation had the following characteristics:

1. The majority of the vascular compliance must reside at the same vascular site in which the filtering of fluid was occurring.
2. The double-clamp technique measured the same P_{pc} as determined by the more laborious isogravimetric procedure.
3. The double-clamp technique actually proved that the capillary pressure of the lung did not change during an isogravimetric procedure, which had previously only been an assumption.
4. The double-clamp method did not require an isogravimetric state or a weighed organ to provide a valid estimate of the pulmonary capillary pressure.

Table 7.1 gives a historical summary of P_{pc} values obtained for normal pulmonary capillary pressures as determined by the isogravimetric technique and other classic methods.

Consequences of Changing R_{pa} and R_{pv} in Lung

Table 7.2 shows how altering R_{pa} and R_{pv} can affect capillary pressure in the lung when $\dot{Q}_B$ is constant, total vascular resistance is either normal (Case 1), increased (Case 2), or decreased (Case 3) and when $R_{pa} = R_{pv}$, $R_{pa} > R_{pv}$ or $R_{pa} < R_{pv}$. When $R_{pa} = R_{pv}$ note that P_{pc} increases above control when total pulmonary vascular resistance increases or decreases (assuming that R_{pa} does not decrease below controls). When $R_{pa} > R_{pv}$, P_{pc} once again increases when total vascular resistance is increased or decreased if $\dot{Q}_B$ remains constant. This

TABLE 7.1. Methods used to estimate pulmonary capillary pressure in canine lungs

Reference	Method	P_{pc} (mm Hg)
Agostoni & Piiper (1962)	Osmometric	8.8
Gaar et al. (1967)	Isogravimetric isolated lung	7.0
Parker et al. (1978)	Intra-alveolar pressure	9.3
Gabel & Drake (1978)	Isogravimetric intact lung	8.7
Bhattacharya & Staub (1980)	Micropuncture	9.8
Linehan et al. (1982); Townsley et al. (1986)	Double occlusion	5.4; 6.0
Parker et al. (1983)	Venous occlusion	7.9
Holloway et al. (1983)	Arterial wedge pressure	6.8
Michel et al. (1984)	Small catheter pressures	9.6
Korthuis et al. (1984)	Modified filtration method	7.3

TABLE 7.2. Pulmonary capillary pressure measurements for variations in the ratio of precapillary to postcapillary resistances when cardiac output = 5 l/min*

	Case	P_{pa}	P_{la}	$P_{pa} - P_{la}$	P_{pc}	R_{pa}	R_{pv}	R_t	P_{pc}
A: $R_{pa} = R_{pv}$	1	20	5	15	12.5	1.5	1.5	3.0	Control
	2	35	5	30	20.0	3.0	3.0	6.0	↑
	3	20	12.5	7.5	16.3	0.75	0.75	1.5	↑
B: $R_{pa} > R_{pv}$	1	20	5	15	10.0	2.0	1.0	3.0	↓
	2	35	5	30	15.0	4.0	2.0	6.0	↑
	3	20	12.5	7.5	15.0	1.0	0.5	1.5	↑
C: $R_{pa} < R_{pv}$	1	20	5	15	15.0	1.0	2.0	3.0	↑
	2	35	5	30	25.0	2.0	4.0	6.0	↑
	3	20	12.5	7.5	17.5	0.5	1.0	1.5	↑

* Modified from Taylor AE, Rehder K, Hyatt B, Parker JC (Eds) *Clinical Respiratory Physiology.* Philadelphia, WB Saunders, 1989.

phenomenon occurs in the decreased resistance state because the vascular pressure drop is smaller across the pulmonary circulation and left atrial pressure increases. In reality, if left atrial pressure (P_{la}) is assumed to remain constant and the drop across the circulation is 7.5, then P_{pa} would be 12.5 and P_{pc} would be: $P_{pc} = 5 + (0.33 \times 7.5) = 7.5$ and P_{pc} would be less than control values. We really don't know what will happen for this condition and it would constitute an important pulmonary circulation study. When $R_{pa} < R_{pv}$, again capillary pressure exceeds the normal P_{pc} value as shown in Case A-1 in Table 7.2. If P_{la} was required to stay constant, then $P_{pc} = 5.0 + (0.66 \times 7.5) = 5.0 + 5.0 = 10.0$, which is only slightly below the control value of 12.5 (Case A-1). Thus, P_{pc} can change unexpectedly in the lung because $\dot{Q}_B$ must remain constant and may even increase under some conditions. Therefore, it is necessary to know P_{pa}, P_{pv}, and $\dot{Q}_B$ to correctly predict P_{pc}.

Description of Pulmonary Circulation Using Four Vascular Segments

Linehan et al. (1982) and Rippe et al. (1987) and our laboratory have extended the concept of a single vascular compliance separated by upstream and downstream resistances as shown in the schematic in Figure 7.1 to a more realistic model as depicted in Figure 7.3A. Accordingly, the pulmonary circulation can be currently described as a resistance-capacitance circuit with a precapillary resistance that is composed of a large arterial segmental resistance (R_{la}) and a small arterial segmental resistance (R_{sa}). The postcapillary resistance is composed of a small venous segmental resistance (R_{sv}) and a large venous segmental resistance (R_{lv}). This circuit as defined by Linehan et al. (1982) and Rippe et al. (1987) has two small equal capacitances located on the arterial (C_{pa}) and venous (C_{pv}) segments of the pulmonary circulation. The third and largest capacitance in the pulmonary circulation is located at the filtration portion of the circulation, (i.e., in the pulmonary capillary segment); otherwise, the double-clamp pressure would not be equal to the isogravimetric capillary pressure measurement.

In order to measure the pressure drops associated with these resistance-capacitance networks, the following experimental techniques were used in an isolated, constant flow, blood perfused dog lung. Figure 7.3B shows the determinations of P_{pao}, P_{pc}, and P_{pvo}, which are the vascular pressures located at the first small capacitance in the arterial segment, the large middle capacitance in the central compartment, and the last small capacitance located in the venous segment of the pulmonary circulation, respectively. First, following rapid occlusion of the pulmonary artery, P_{pa} is seen to drop rapidly to P_{pao}. After unclamping the arterial line, the venous segment was suddenly clamped and P_{pv} is seen to rapidly rise to P_{pvo}. Again, after reestablishing blood flow the double occlusion technique was used to measure P_{pc}. Since each segmental vascular pressure is known, a resistance can be calculated for each vascular segment. This technique in reality measures an inductance composed of both resistance and compliance (capacitance) in each vascular segment; however, since the arterial and venous compliances (capacitances) are very small, these measurements provide a good estimate of the resistance in each vascular segment. Table 7.3 shows measurements of each of these pressures in dog lungs at control conditions, and during infusions of histamine, serotonin, and norepinephrine. When flow was constant at normal cardiac output conditions, capillary pressure increased during the infusion of all three compounds, yet histamine's effect was predominantly postcapillary, serotonin's effect was mainly precapillary, and norepinephrine had a greater effect on postcapillary resistance than precapillary resistance.

Also shown in Table 7.3 are the pulmonary vascular segmental compliances calculated at control conditions and for the various drug treatments from the equations derived by Linehan et al. (1982). For total pulmonary vascular

A.

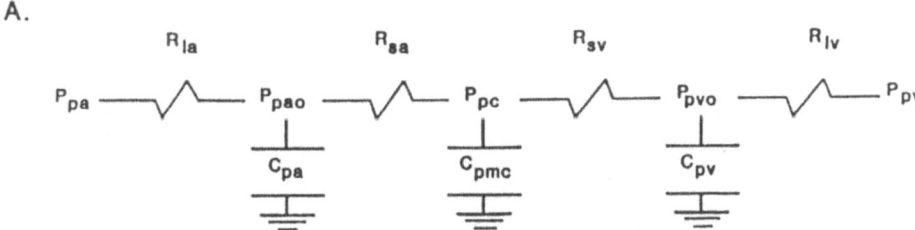

B.

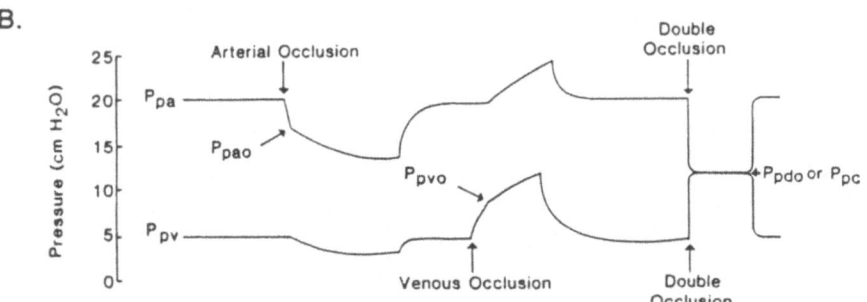

FIGURE 7.3. A: Schematic representation of the longitudinal distribution of vascular resistances and compliances in the canine pulmonary circulation. R_{la}, large artery resistance; R_{sa}, small artery resistance; R_{sv}, small vein resistance; R_{lv}, large vein resistance; P_{pa}, pulmonary artery pressure; P_{pao}, arterial occlusion pressure; P_{pc}, pulmonary capillary pressure; P_{pvo}, venous occlusion pressure; P_{pv}, pulmonary vein pressure; C_{pa}, arterial compliance; C_{pmc}, middle compartment compliance; C_{pv}, venous compliance. B: Pressure tracings showing arterial occlusion pressure (P_{pao}), venous occlusion pressure (P_{pvo}), and double-occlusion pressure (P_{pdo} or P_{pc}).

compliance:

$$C_{pt} = \dot{Q}/(\Delta P/\Delta t) \qquad (7.4)$$

where total pulmonary vascular compliance (C_{pt}) was calculated from the relationship of the slope of the venous pressure-time transient ($\Delta P/\Delta t$) obtained by venous occlusion and blood flow ($\dot{Q}$). Middle compartment vascular compliance was calculated from the following equations:

$$C_{pmc}/C_{pt} = [4(P_{pai} - P_{pv})/(P_{pa} - P_{pv}) - 0.75)]^{1/2} \qquad (7.5)$$

where P_{pai} is the arterial pressure obtained by extrapolating the slope of the arterial pressure rise that occurs with venous occlusion back to zero time. When total compliance and middle compartment compliance are known,

TABLE 7.3.   Pulmonary segmental vascular pressures, vascular resistances and vascular compliances in canine lungs*

	Pulmonary segmental vascular pressures (cm H_2O)				
	P_{pa}	P_{pao}	P_{pc}	P_{pvo}	P_{pv}
Control	16.0	12.1	10.1	8.0	5.0
Histamine	41.0	36.2	31.0	28.3	3.3
Serotonin	46.5	32.4	20.7	14.5	3.0
Norepinephrine	34.3	25.3	19.7	16.6	3.3

	Pulmonary segmental vascular resistances (cm $H_2O/1/min/100$ g)					
	R_{la}	R_{sa}	R_{sv}	R_{lv}	R_{pa}/R_{pv}	R_{pt}
Control	4.0	2.0	0.5	4.0	1.33	10.5
Histamine	4.8	4.9	17.5	23.0	0.24	50.2
Serotonin	24.0	20.0	10.0	20.0	1.47	74.0
Norepinephrine	17.0	10.0	6.0	25.0	0.87	58.0

	Compartmental pulmonary vascular compliances (ml/cm $H_2O/100$ g)		
	C_{pt}	C_{pmc}	$C_{pa} + C_{pv}$
Control	1.30	1.15	0.15
Histamine	0.39	0.35	0.04
Serotonin	0.67	0.37	0.30
Norepinephrine	0.85	0.75	0.10

*From Rippe et al. (1987).

large vessel compliance can be obtained from the following relationship:

$$C_{pa} + C_{pv} = C_{pt} - C_{pmc} \qquad (7.6)$$

where C_{pa} and C_{pv} are the large vessel compliances, respectively. In Table 7.3 note that the middle compartment compliance is normally 90% of the total vascular compliance. This relationship does not change with histamine or norepinephrine, but serotonin reduces the middle compartment vascular compliance to approximately 55% of the total lung compliance. From the preceding discussions, it is obvious that capillary pressures and segmental resistances can be measured in experimental conditions that are now being studied in several laboratories throughout the world. But, how does this measurement relate to the physiological functions of the lung?

Importance of P_{pc} in Determining Lung Fluid Balance

Figure 7.4 shows classic data from Guyton and Lindsey (1959) wherein the rate of lung edema formation was plotted as a function of left atrial pressure (or capillary pressure). The data depicted by the curve on the far right indicate that almost no fluid enters the lung tissue until left atrial pressure exceeds 25 mm Hg. As left atrial pressure further increases, fluid begins to enter the

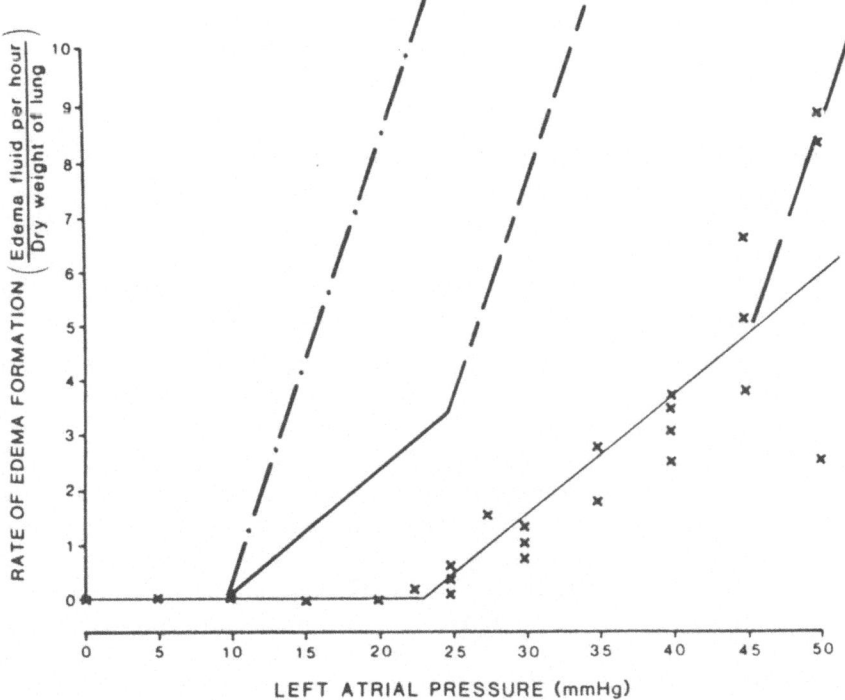

FIGURE 7.4. Plot of lung edema formation as a function of pulmonary capillary pressure for normal capillary pressure (*right curve*), damaged capillaries (*middle curve*), and damaged capillaries with accelerated weight gain (*left curve*). (Modified from Guyton and Lindsey, 1959, by permission of the American Heart Association Inc.

tissues, and at very high left atrial pressures the weight gain appears to accelerate. Thus, lungs do not usually become edematous until left atrial pressures exceed 25–30 mm Hg. However, if the endothelium is damaged by some pathological process, the plasma proteins are not as effective in holding fluid in the circulation; therefore, fluid will begin to enter the lungs at much lower capillary pressures (shown by the middle curve in Figure 7.4). In fact, when the lung is even more susceptible to edema, small changes in left atrial pressures above normal can cause excessive accumulations of lung water as shown by the dashed-dotted line of the left-hand curve of Figure 7.4.

Over the years, we have discussed how the lungs can resist the formation of pulmonary edema when capillary pressure is elevated in terms of "safety factors against edema." These can best be described in relation to the modified Starling equation:

$$J_v = K_{fc}[(P_{pc} - P_t) - \sigma_d(\pi_p - \pi_t)] \tag{7.7}$$
$$\underset{\text{pressure}}{\text{filtration}} \qquad \underset{\text{pressure}}{\text{absorption}}$$

which describes the volume flow that occurs into the tissue (J_v) in terms of capillary pressure (P_{pc}) and tissue pressure (P_t), and the colloid osmotic pressure of plasma (π_p) and tissues (π_t). Normally, these factors operate to oppose edema by the following mechanisms:

1. Lymph flow increases, which removes some of the fluid that is filtered across the capillary wall as capillary pressure is increased.
2. Interstitial fluid pressure increases, which decreases the hydrostatic filtration pressure gradient ($P_{pc} - P_t$).
3. The colloid osmotic absorption pressure increases, which is associated with decreasing π_t as the capillaries filter a protein-poor fluid into the tissues.

These factors considered together consist of the "edema safety factor," which equals approximately 20 mm Hg in normal lungs. When the capillaries are damaged, the absorption pressure decreases because π_t increases and the reflection coefficient, σ_d (which is a measure of the permeability of the capillary wall to proteins) also decreases, causing the absorption pressure to decrease in a multiplicative fashion rather than in a simple additive fashion. The filtration coefficient (K_{fc}) increases, which will cause a greater filtration to occur across the capillary wall for a given difference in the hydrostatic filtration and colloid osmotic absorption forces. Therefore, it is imperative that the physician (or pulmonary physiologist) know the value of P_{pc} in order to describe the actual fluid balance status of the lung. P_{pc} can be used as a means of properly assessing any fluid and drug therapy employed to prevent alveolar edema.

Determination of P_{pc} in In Situ Animal Lungs and in Human Lungs

Holloway et al. (1983) demonstrated that capillary pressures could also be estimated in close-chested dogs by using a Swan-Ganz catheter placed into either the pulmonary artery or pulmonary vein. Figure 7.5 shows tracings of arterial and venous pressures in dog lungs in which the balloons on the Swan-Ganz catheters were inflated either in the pulmonary artery or pulmonary vein of the closed-chested dogs. When the chart speed was slow, the pressures showed inflection points that are equal to P_{pc}. When these values were compared to the classic Gaar equation, they agreed surprisingly well. In addition, when these curves were analyzed mathematically by assuming that the central capacitance discharges exponentially to the pulmonary arterial wedge measurements, both estimates of P_{pc} were identical. In earlier studies, Parker et al. (1983) and Ryan et al. (1980) had shown that the inflection points of both the arterial and venous clamp pressures seen in the slow-recorded arterial and venous pressure transients approximated isogravimetrically determined capillary pressures in isolated lungs. Therefore, it was thought that these pressures would be useful estimates of pulmonary capillary pressures in

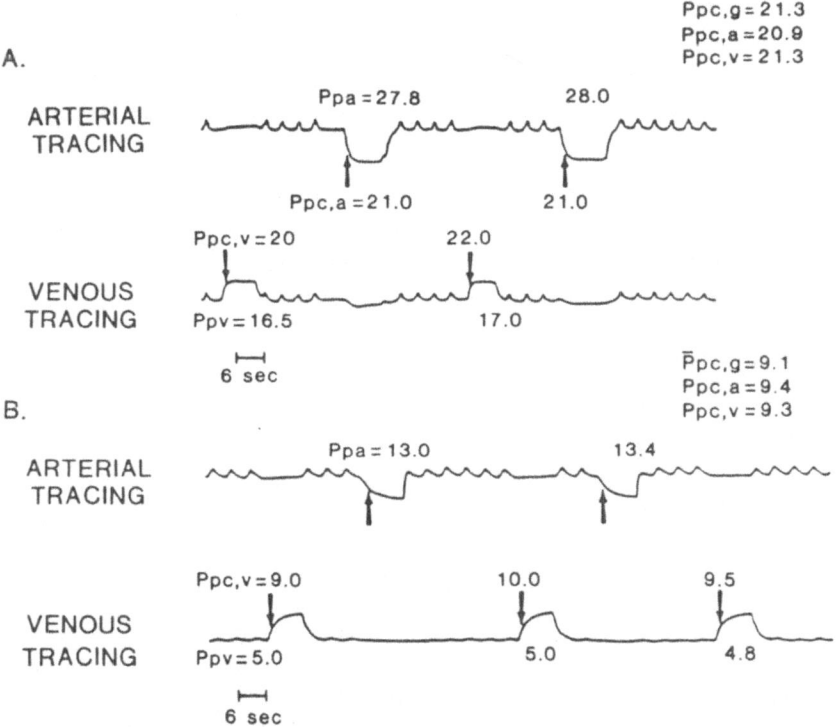

FIGURE 7.5. Tracings of arterial and venous pressures during alternating arterial and venous vascular occlusions with Swan-Ganz balloon catheters. *A*: Recordings obtained at elevated pulmonary vascular pressures. *B*: Tracings obtained at normal pulmonary vascular pressures. $P_{pc,g}$, capillary pressure calculated using Gaar equation; $P_{pc,a}$, capillary pressure calculated from arterial occlusion; $P_{pc,v}$, capillary pressure estimated from venous occlusion, respectively. (Modified from Holloway et al., 1983.)

human lungs. Figure 7.6 shows the rationale in using the arterial clamp measurements to estimate P_{pc}. The upper curve (Fig. 7.6A) shows an easy way of determining the inflection point following wedging of the pulmonary arterial catheter. A line is drawn on the rapid portion of the curve and another line is drawn on the slow transient. The point where the two lines cross was shown by Holloway et al. (1983) to be equal to that determined by plotting the log of $P_{pa} - P_{p,wedge}$ as a function of time and then extrapolating to zero to obtain P_{pc} (as seen in Fig. 7.7B, below). Figure 7.6B shows that it is very easy to obtain information as to whether the majority of the pulmonary resistance is pre- or postcapillary. If the majority of the pulmonary resistance is located upstream from the capillaries, then when the pulmonary arterial catheter is wedged, P_{pa} will drop very rapidly towards $P_{p,wedge}$ (solid line in Fig. 7.6B). Since the inflection point occurs near the wedge pressure, then the capillary pressure is almost equal to the wedge measurement. However, when

A.

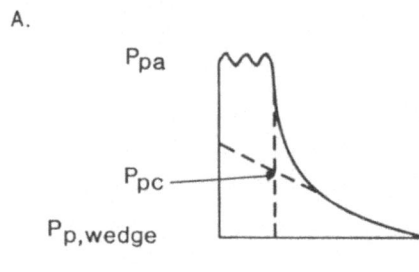

B.

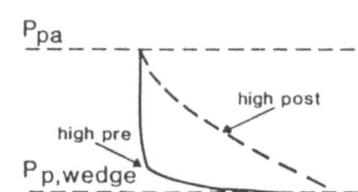

FIGURE 7.6. *A*: Plot of P_{pa} and $P_{p,wedge}$ following arterial occlusion. The two curves are extrapolated until they intersect, which estimates capillary pressure (P_{pc}). *B*: Plot of P_{pa} and $P_{p,wedge}$ tracings following arterial occlusion. When the precapillary resistance is high, P_{pa} falls rapidly toward $P_{p,wedge}$. When postcapillary resistance is high, P_{pa} falls more slowly toward $P_{p,wedge}$.

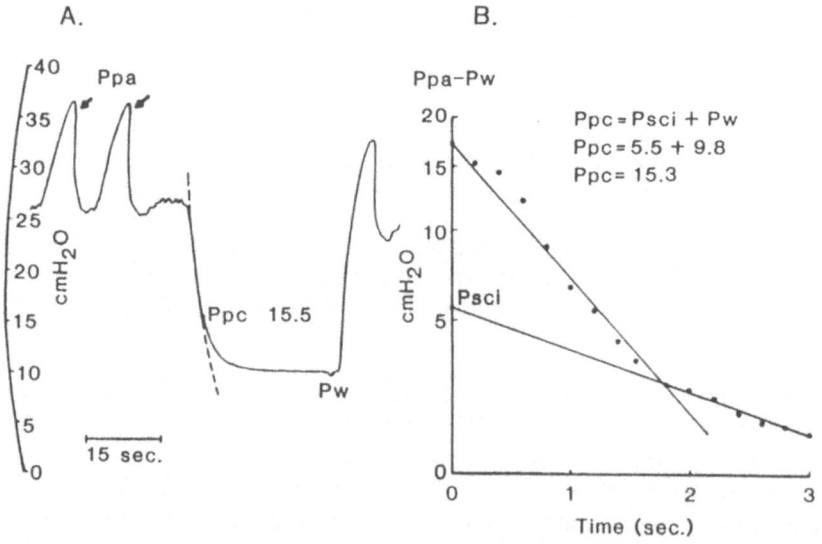

FIGURE 7.7. *A*: Pressure recording on curvilinear paper before and after balloon occlusion of a canine pulmonary artery. The *dashed line* represents the point at which the pressure tracing deviates from the curved template that estimates capillary pressure (P_{pc}). *B*: If the difference between P_{pa} and P_w at 0.2-s intervals after pulmonary artery occlusion is plotted on a semilogarithmic scale as a function of time, the slow linear component can be extrapolated back to the time of occlusion. The pressure at the slow-component intercept (P_{sci}) is added to P_w to obtain P_{pc}. This calculation yielded a P_{pc} of 15.3 cm H_2O for the pressure tracing shown in *A*. (Modified from Cope et al., 1986 copyright © 1986 by Williams & Wilkins.)

the resistance is located predominantly in the postcapillary segment of the circulation, P_{pa} slowly approaches $P_{p,wedge}$. Although the capillary pressure in this condition is more difficult to ascertain, the resistance is obviously on the venous side. Cope et al. (1986) have recently applied this measurement of capillary pressure to the patient bedside, and Figure 7.7 shows a measurement of P_{pc} conducted in human lungs. In the left-hand panel, a line was drawn on the rapid component of the curve to estimate the inflection point of the curve following inflation of the Swan-Ganz catheter balloon. The right-hand panel shows a plot of $P_{pa} - P_{p,wedge}$ for the pressure transient. When the slow component was extrapolated to zero time, P_{pc} was calculated from the sum of the wedge and the pressure intercept (P_{sci}), which was 15.3 cm H_2O. This more mathematical measure of P_{pc} agrees well with the estimated value of 15.5 obtained using the inflection point of the arterial pressure transient.

Table 7.4 shows average values of pulmonary capillary pressures obtained by Cope et al. in several cardiac patients undergoing surgery to replace either aortic or mitral valves, and before and following bypass surgery. The ratio of P_{pc} to $P_{p,wedge}$ in the pulmonary circulation was fairly constant before and after surgery in both the coronary bypass and aortic valve replacement patients, although cardiac output was improved in the bypass patients, which indicated a decreased total pulmonary vascular resistance. However, P_{pc} changed significantly in the postoperation mitral valve replacement patients, who had an increased total resistance that was located predominantly in the venous segment of their lungs' circulation. Therefore, it appears that these patients have high pulmonary vascular tone due to the fact that when the higher wedge pressures are reduced the veins constrict, which results in a higher capillary pressure due to the increased postcapillary resistance.

TABLE 7.4.   Pulmonary hemodynamic measurements in humans*

	P_{pa} (mm Hg)	$P_{p,wedge}$ (mm Hg)	P_{pc} (mm Hg)	$P_{pc}/P_{p,wedge}$ (mm Hg)	CO** (l/min)	R_{pa}/R_{pv} (dynes/s/cm^5)
Coronary Artery Bypass						
Preop	22.6	16.6	18.3	1.11	4.52	2.81
Postop	24.4	17.2	19.2	1.14	6.50	2.73
Aortic Valve Replacement						
Preop	27.4	20.7	22.7	1.11	6.44	3.00
Postop	23.2	15.3	17.7	1.16	6.70	3.15
Mitral Valve Replacement						
Preop	35.2	24.3	26.8	1.10	4.72	3.36
Postop	39.6	22.1	28.2	1.28	4.82	1.87

*Modified from Cope et al. Capillary pressure measurement in human lungs. In *Proceedings of the 3rd International Conference on ARDS.* (in press).
**CO, cardiac

Conclusion

Obviously, the measurement of pulmonary vascular segmental resistances and compliances will be useful in determining how various vasoactive substances affect the pulmonary circulation, but more importantly, they will be extremely useful in determining the prevailing capillary pressure—the most important force that regulates pulmonary fluid balance. Although the studies conducted in human lungs are in the preliminary stages, there is no doubt that these measurements will yield pertinent information concerning the lung fluid balance state of the patient. Finally, it must be emphasized that these measurements of segmental resistances and compliances will certainly lead to the collection and dissemination of new information relating pulmonary capillary pressure to lung fluid balance and pulmonary hemodynamics. The studies of Hakim, Dawson, Linehan, and numerous other investigators have opened this new vista, and there is still a great deal of information to be learned about the pulmonary circulation using the approaches presented in this book.

References

Agostoni E, Piiper J (1962) Capillary pressure and distribution of vascular resistance in isolated lung. *Am J Physiol* 202:1033–1036.

Bhattacharya J, Staub N (1980) Direct measurement of microvascular pressures in the isolated dog lung. *Science* 210:327–328.

Cope D, Allison RC, Parmentier JL, Miller JN, Taylor AE (1986) Measurement of effective pulmonary capillary pressure using the pressure profile after pulmonary artery occlusion. *Crit Care Med* 14:16–21.

Dawson CA, Linehan JH, Rickaby DA (1982) Pulmonary microcirculatory hemodynamics. *Ann NY Acad Sci* 384:90–106.

Gaar KA Jr, Taylor AE, Owens LJ, Guyton AC (1967) Pulmonary capillary pressure and filtration coefficient in the isolated perfused lung. *Am J Physiol* 213:910–914.

Gabel JC, Drake RE (1978) Pulmonary capillary pressure in intact dog lungs. *Am J Physiol* 235:H569–H573.

Guyton AC, Lindsey AW (1959) Effect of elevated left atrial pressure and decreased plasma protein concentration on the development of pulmonary edema. *Circ Res* 7:689–693.

Hakim TS, Dawson CA, Linehan JH (1979) Hemodynamic responses of dog lung lobe to lobar venous occlusion. *J Appl Physiol* 47:145–152.

Holloway H, Perry M, Downey J, Parker J, Taylor A (1983) Estimation of effective pulmonary capillary pressure in intact lungs. *J Appl Physiol* 54:846–851.

Korthuis RJ, Townsley MI, Rippe B, Taylor AE (1984) Estimation of isogravimetric capillary pressure by a filtration method in skeletal muscle and lung. *J Appl Physiol* 57:1817–1823.

Linehan JH, Dawson CA, Rickaby DA (1982) Distribution of vascular resistance and compliance in a dog lung lobe. *J Appl Physiol* 53:158–168.

Michel RP, Hakim TS, Chang HK (1984) Pulmonary arterial and venous pressures measured with small catheters in dogs. *J Appl Physiol* 57:309–314.

Pappenheimer JR, Soto-Rivera A (1948) Effective osmotic pressure of the plasma proteins and other quantities associated with the capillary circulation in the hind-limbs of cats and dogs. *Am J Physiol* 152:471–491.

Parker JC, Guyton AC, Taylor AE (1978) Pulmonary interstitial and capillary pressures estimated from intra-alveolar fluid pressures. *J Appl Physiol* 44:267–276.

Parker JC, Kvietys PR, Ryan KP, Taylor AE (1983) Comparison of isogravimetric and venous occlusion capillary pressures in isolated dog lungs. *J Appl Physiol* 55:964–968.

Rippe B, Parker JC, Townsley MI, Mortillaro NA, Taylor AE (1987) Segmental vascular resistances and compliances in dog lung. *J Appl Physiol* 62:1206–1215.

Ryan K, Kvietys P, Parker JC, Taylor AE (1980) Comparison of venous occlusion and isogravimetric capillary pressure in isolated dog lungs. *Physiologist* 23:76.

Townsley MI, Korthuis RJ, Rippe B, Parker JC, Taylor AE (1986) Validation of double vascular occlusion method for $P_{c,i}$ in lung and skeletal muscle. *J Appl Physiol* 61:127–132.

8
Sites of Vasoactivity in the Pulmonary Circulation Evaluated Using Rapid Occlusion Methods

JOHN H. LINEHAN, CHRISTOPHER A. DAWSON, THOMAS A. BRONIKOWSKI, and DAVID A. RICKABY

Introduction

The transient pressure and flow data obtained following rapid occlusion of the arterial inflow and/or venous outflow from a lung lobe contain information about the arterial-to-venous distribution of vascular resistance (R) relative to the distribution of vascular compliance (C) (Bronikowski et al., 1985; Bshouty et al., 1987; Dawson et al., 1982; Hakim et al., 1979, 1982, 1983; Holloway et al., 1983; Rippe et al., 1987; Rock et al., 1985). Interest in this approach has, to a large extent, centered around the potential for following changes in microvascular pressure and determining the arterial or venous site of action of vasomotor stimuli. In the following discussion we will present some of our ideas on the information content of occlusion data as interpreted using mathematical models in an attempt to provide useful parameters descriptive of the pulmonary microvascular bed.

The Simple Resistance-Compliance Model

Following rapid occlusion of the venous outflow from an isolated dog lung lobe perfused with constant flow, the venous pressure jumps quickly to a pressure somewhere between the preocclusion arterial and venous pressures. Subsequently, the vascular bed fills with blood at the constant inflow rate and the arterial and venous pressures begin to rise more slowly. An example of the pressure versus time from occlusion curves is shown in Figure 8.1 (Linehan et al., 1982). A simple model that provides useful insight for interpreting the rapid jump in venous pressure in response to venous occlusion is a hemodynamic resistance–vascular compliance (RC) model in which the lobar vascular bed is represented by two series resistances, one upstream and one downstream of a central vascular compliance. The electrical equivalent of this model is a RC single T section shown in Figure 8.2. In this model, when flow through the downstream (venous) resistance (R_v) stops, the venous pressure rises immediately to the pressure of the centrally located compliance (C_L). As

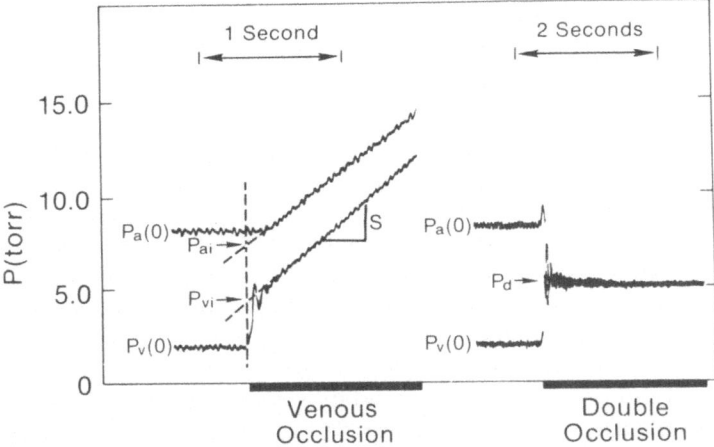

FIGURE 8.1. Arterial and venous pressure tracings obtained during venous occlusion and double-occlusion experiments on a dog lung lobe. P_{ai} and P_{vi} are obtained by extrapolation of $P_a(t)$ and $P_v(t)$ to the instant occlusion. P_d is the equilibrium pressure following double occlusion. S, the slope of the $P_v(t)$, is inversely proportional lobar vascular compliance. (Adapted from Linehan et al., 1982.)

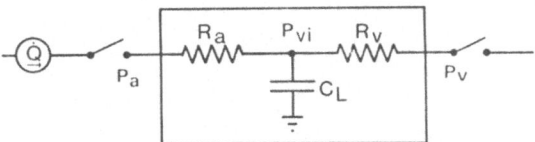

FIGURE 8.2. An electrical analogue T section used to explain the features of the arterial and venous pressure responses to occlusion of the venous outflow, arterial inflow, or both.

flow continues into the pulmonary artery, the resulting increase in lobar volume is stored in the compliance. This model predicts that both arterial (P_a) and venous pressure (P_v) rise in a parallel fashion, as in Figure 8.1, and that the post-occlusion pressure difference, $P_a - P_v$, is due to the upstream (arterial) resistance (R_a). Thus, this black box model provided a simple and objective means for identifying a pressure (P_{vi}), associated with the compliant region. In practice, the slowly rising portion of the venous pressure curve could be linearly extrapolated back to the instant of venous occlusion to obtain P_{vi} as shown in Figure 8.1. Several lines of evidence suggested that P_{vi} was a useful microvascular pressure in the dog lung lobe. For example, when flow ($\dot{Q}$) is held constant, the upstream and downstream pressure drops ($P_a - P_{vi}$ and $P_{vi} - P_v$) could be changed by various pulmonary vasomotor stimuli paralleling the changes defined using the low-viscosity bolus technique (Hakim et al., 1979). This comparison of results suggested that constriction of the lobar arteries increased the pressure drop across R_a, the upstream resistance, while constriction of the lobar veins increased the pressure drop across R_v, the downstream resistance.

$$R_a = \frac{P_a(0) - P_{vi}}{\dot{Q}} \quad \text{and} \quad R_v = \frac{P_{vi} - P_v(0)}{\dot{Q}} \tag{8.1}$$

where 0 refers to the time of occlusion.

The isogravimetric method, as applied to the lung by Gaar et al. (1967), has generally been accepted as measuring the effective microvascular pressure (sometimes referred to as the pressure at the filtration midpoint) in isolated lungs. The observation that P_{vi} is close to the isogravimetric pressure (Dawson et al., 1982; Parker et al., 1983) suggests that the microvascular hydrostatic pressure controlling transvascular fluid movement and the pressure controlling lobar vascular volume (P_{vi} in Fig. 8.1) are nearly the same. The simple model used to explain the arterial and venous pressure curves following venous occlusion also predicts certain results of other occlusion experiments. For example, when both the arterial inflow and the venous outflow are simultaneously occluded (double occlusion), the instantaneous blood volume is trapped in the lobe. According to the model in Figure 8.2, P_a instantly decreases and P_v instantly increases to a common equilibrium pressure (referred to subsequently as P_d). Experimental comparison of the results of double and venous occlusions (i.e., P_d and P_{vi}, respectively) showed that P_d was in close agreement with P_{vi} (Dawson et al., 1982).

Multicompartment RC Models

The simple RC model appears to be useful for identifying the arterial and venous site of action of pulmonary vasomotor stimuli and for obtaining a reasonable estimate of the microvascular pressure under conditions in which the normal symmetry of the arteriovenous resistance distribution in the lungs has been altered, for example, by arterial or venous constriction. However, the simple model does not account for all of the features of the occlusion data, suggesting that a more complex model may be useful. Adding complexity to the model has at least two goals. The first is to identify quantitative bounds on conclusions drawn using the simple model to evaluate the data. The second is to determine whether a model that takes advantage of additional features of the data can be used to gain more insight into the mechanics of the pulmonary circulation. Although the simple model reproduces the most striking features of the occlusion data (e.g., the rapid rise in venous pressure following venous occlusion, followed by a slower rise in both arterial and venous pressures), several aspects of the data point to the fact that the resistance and compliance are distributed from lobar artery to lobar vein throughout the vascular bed. For example, there is a short time delay before the arterial pressure begins to rise following venous occlusion. This time delay is not predicted by the simple RC model. In addition, we found that P_d is consistently a little higher than P_{vi} (Dawson et al., 1982).

These distinctions between the simple model predictions and the data suggest the possibility of describing the vascular resistance versus compliance distribution in the lung lobe in more detail. In examining this possibility, we found that there were more distributed compartmental RC networks that could be used to take advantage of an additional piece of information obtainable from the venous occlusion data, namely, the time delay in the arterial pressure curve. This time delay can be converted into the intercept pressure, P_{ai}, by extrapolating the arterial pressure curve back to the instant of occlusion as indicated in Figure 8.1 (Linehan et al., 1982). These models with a small number of compartments allowed for the calculation of specific compartmental parameters. They also suggested that, even though the vascular resistance and compliance are distributed, there is a region between muscular arteries and veins in the normal dog lung lobe that includes a larger fraction of the vascular compliance than of the vascular resistance. This characteristic of the intralobar vascular bed is the basis for the determination of the arterial and/or venous site of vaso-constriction and the estimation of microvascular pressure, P_c, using the occlusion methods (Hakim et al., 1979; Linehan and Dawson, 1983; Linehan et al., 1982).

The Continuous Resistance Versus Compliance Distribution

One problem with using a small number of compartmental elements to represent the actually continuous distribution of intralobar R and C is that the relationships between anatomic and model compartments are not specified. We have attempted to deal with this problem by examining the theoretical limits of what can be learned about the actual continuous distribution using the venous occlusion data (Bronikowski et al., 1984, 1985). To this end, we visualized the serial or longitudinal distributions of the local vascular resistance $R(x)$ and compliance $C(x)$ as functions of a normalized spatial variable (x) that increases from $x = 0$ at the lobar artery inlet to $x = 1$ at the venous outlet. The x variable might be, for example, the fractional distance from the arterial inlet to the venous outlet or fractional cumulative vascular volume from the inlet to the outlet. Local vascular resistances and compliances tend to be distributed differently with respect to the spatial variable as indicated by the structure of the occlusion data previously discussed in regard to the compartmental models. We have found it useful to view the relationship between R and C in terms of cumulative functions. For example, R_{cum} means the sum of the vascular resistances between $x = 0$ and any x. $R_{cum}(1)$ is then equal to the total lobar vascular resistance. The utility of this approach can be exemplified using the P_{ai} and P_{vi} obtainable from the venous occlusion data to put bounds on the possible graphs of the cumulative R versus C functions. This can be done by representing the continuously distributed vascular bed by an arbitrarily large number of serial resistances and parallel compliances. When this is done, it can be shown that the intercept pressures from the venous occlusion data define specific boundaries containing the continuous cumulative R versus C distribution as shown schematically as region W in Figure 8.3. That is, all continuous R_{cum} versus C_{cum} graphs compatible with the steady state pressures, $P_a(0)$ and $P_v(0)$, and pressures P_{ai} and P_{vi} are continuously increasing functions within the region W. One important result of this analysis of the continuous model is that the measureable pressure data can be thought of as placing secure bounds on the value of P_c, which is now defined as the vascular pressure at the midpoint of the cumulative vascular compliance, as shown in Figure 8.4. P_c, as the mean of the bounds, is defined by:

$$P_c(0) - P_v(0) = (P_a(0) - P_{ai}) + (P_{vi} - P_v(0)) \qquad (8.2)$$

The grapical interpretation of this equation can be inferred by examining Figure 8.4. When $P_a(0) - P_{ai}$ is a small fraction of $P_a(0) - P_v(0)$, the bounds on P_c are narrow. This occurs when most of the vascular compliance and a relatively small fraction of the vascular resistance are located in the microvascular bed, which is again a key feature of the pulmonary vascular bed that accounts for the utility of the occlusion methods.

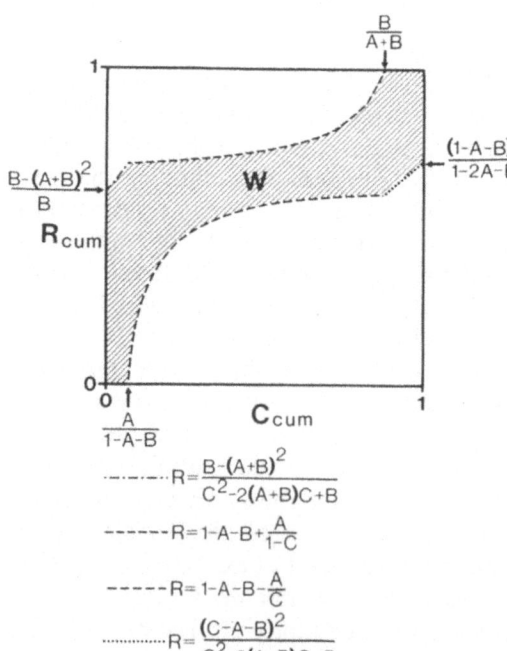

FIGURE 8.3. The region W in the cumulative RC plane contains all cumulative R versus C distributions that are compatible with P_{ai} and P_{vi} data. The boundaries of W are in terms of A and B where $A = (P_a(0) - P_{ai})/(P_a(0) - P_v(0))$ and $B = (P_{vi} - P_v(0))/(P_a(0) - P_v(0))$. In the equations for the boundaries of W, R_{cum} and C_{cum} are abbreviated by R and C, respectively. (Adapted from Bronikowski et al., 1984.)

FIGURE 8.4. The region defined according to Figure 8.3 with P_c located at $C_{cum} = 0.5$. P_c is bounded such that $B - A \leq (P_c - P_v(0))/(P_a(0) - P_v(0)) \leq B + 3A$. (Adapted from Bronikowski et al., 1985.)

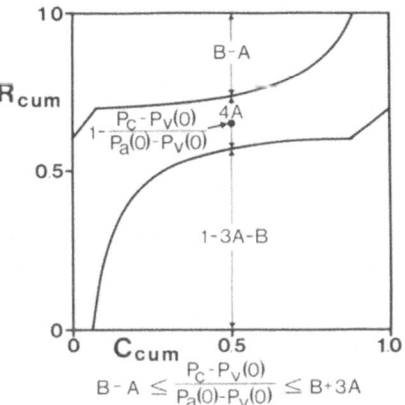

$$B - A \leq \frac{P_c - P_v(0)}{P_a(0) - P_v(0)} \leq B + 3A$$

Biomechanically Based Models

The previously described mathematical models are "black box" models in the sense that the structural and functional relationships implied by the model elements are not accessible to direct observation. Fung (1984) and Zhuang et al. (1983) have presented a deterministic model of the steady state (constant flow) hemodynamics of the pulmonary circulation of the cat that is based on direct measurements of vessel morphometry and elasticity. This deterministic model offers a unique opportunity to compare the predictions of the black box models with simulations obtained from a hemodynamic model based on measured structural and physical properties of the vascular bed. Such comparisons should help to put the various assumptions involved in the black box analyses and in the deterministic modeling into perspective. We have extended the original steady state model of Zhuang et al. (1983) to include vascular compliance (Krishnan et al., 1986) and both vascular compliance and blood inertance (Linehan et al., 1988) so that it can reproduce some of the transient phenomenon that may be important in evaluating the occlusion responses. We call these models "dynamic models."

For these dynamic models we have again used a compartmental approach. The compartments are diagrammatically represented by the electrical analogy in Figure 8.5. For the cat lung, each compartment represents one of 11 orders each of arteries and veins connected by the capillary sheet (Zhuang et al., 1983). The values of resistance (R) compliance (C) and inertance (L) were assigned to each order using the cat lung morphometric and distensibility data of Zhuang et al. (1983) with small adjustments to account for differences in the total hemodynamic resistance and compliance and the longitudinal distributions of resistance and compliance between individual cat lungs (Krishnan et al., 1986).

For the kth order of vessels in the model, the following system of differential equations represents the momentum and volume balances, including the

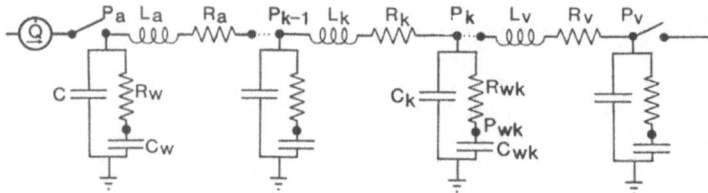

FIGURE 8.5. The analogue representation of the cat lung dynamic model. P_a and P_v represent the pulmonary artery and vein pressures. For steady flow, both switches are closed. In the analogy, the distal switch opens for venous occlusion, the proximal switch opens for arterial occlusion, and both switches open for double occlusion. (Adapted with permission from ASME/BED *Adv in Bioeng*, Linehan et al., 1988.)

vessel wall viscoelasticity (Linehan et al., 1988)

$$P_{k-1} - P_k = L_k \frac{d\dot{Q}_k}{dt} + R_k \dot{Q}_k \tag{8.3}$$

$$\dot{Q}_k - \dot{Q}_{k+1} = C_k \frac{dP_k}{dt} + \frac{P_k - P_{wk}}{R_{wk}} \tag{8.4}$$

$$\frac{P_k - P_{wk}}{R_{wk}} = C_{wk} \frac{dP_{wk}}{dt} \tag{8.5}$$

For the kth order vessels, $P_{k-1} - P_k$ and $\dot{Q}_k$ are the instantaneous pressure difference and flow, and R_k and L_k are the hemodynamic resistance and inertance. C_k and C_{wk} are the vascular compliances, and R_{wk} is the viscous wall resistance of the St. Venant body (Milnor, 1982) used in this model to represent the possible effects of wall viscoelasticity. Symmetry, in regard to the order of the RCL elements, is accomplished by placing half of the capillary sheet compliance both proximal and distal to capillary hemodynamic resistance (Fig. 8.5). The R, L, and C of each order are functionally related to vessel diameter (and, hence, vessel volume). In addition, since vessel volume changes during the occlusion-induced transients, the R, L, and C are variable during the transient.

The following formulas were used to calculate the values of R, L, and C for each order. To calculate vascular resistance, Poiseuille's formula was used. For each order of arteries and veins,

$$R = \frac{128\mu l}{N\pi D^4} \tag{8.6}$$

and for the capillary sheet,

$$R = \frac{12\mu f l^2}{(VSTR)Ah^3} \tag{8.7}$$

where μ is the blood viscosity of the particular order, N is the number of vessels in the order, l is the vessel or sheet length, D is the vessel diameter based on the mean vessel pressure, h is the capillary sheet thickness based on the mean sheet pressure, f is the friction factor (1.6 from Fung, 1984), A is the sheet area (0.84 m^2 for the simulations described below), and VSTR is the ratio of blood volume to total volume within the sheet (0.916 as indicated by Zhuang et al., 1983). Eqs. 8.6 and 8.7 were used in lieu of the formulas given by Fung et al. (Fung, 1984; Zhuang et al., 1983) that account for vessel dimension variation with pressure along the vessel length. A comparison of steady state results using the formulas given by Fung (1984) and Eqs. 8.6 and 8.7 indicated little difference, and the simplifications offered by using formulas based on a uniform vessel diameter are advantageous for formulating the dynamic model. For a uniform vessel diameter, the fluid inertance, L, is calculated from

$$L = \frac{4\rho l}{N\pi D^2} \tag{8.8}$$

where ρ is the blood density. The fluid inertance of the capillary sheet is negligible, and not included in the model.

Static vascular compliance, C_s, is defined as

$$C_s = C + C_w = \frac{dQ}{dP_t} \tag{8.9}$$

where P_t is the mean transmural pressure, equal to the vascular pressure–pleural pressure difference for orders 5–11, and to the vascular pressure–alveolar pressure difference for orders 1–4 and the capillary sheet. The Q is the vascular volume. For the arteries and veins,

$$Q = \frac{N\pi D^2 l}{4} \tag{8.10a}$$

and for the capillary sheet

$$Q = Ah(\text{VSTR}) \tag{8.10b}$$

We found (Krishnan et al., 1986) that the dynamic compliance observable from the slope (S in Fig. 8.2) of the pressure tracings following venous occlusion could be modeled by assuming C was approximately equal to C_w, and this assumption allowed assignment of the parameter values within the St. Venant elements. Assuming constant pleural and alveolar pressures, changes in vessel transmural pressure equal changes in intravascular pressure. Fung et al. (1984) found a linear relationship for the dependence of D on P_t over a limited range of P_t. However, as P_t increases beyond this range, dD/dP_t decreases and the vessels approach their distensibility limits. The result is that vascular compliance decreases with increasing luminal pressure. We found it convenient to represent the function $D(P_t)$ in the R, L, and C formulas by the following exponential

$$D(P_t) = \gamma D_0 - (\gamma - 1)D_0 e^{-(\beta_0 P_t)} \tag{8.11}$$

where D_0 is the vessel diameter when $P_t = 0$ and β_0 is the vessel compliance factor (Zhuang et al., 1983). The γ is defined to be the ratio $D_{\max}/D_0$ where

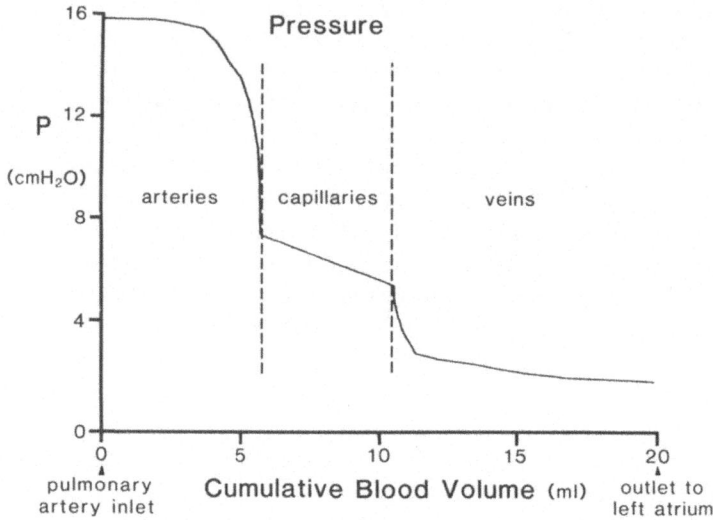

FIGURE 8.6. The simulated steady state pressure distribution within the cat lung plotted as a function of the cumulative blood volume. (Adapted with permission from ASME/BED *Adv in Bioeng*, Linehan et al., 1988.)

D_{max} is a limiting vessel diameter at large values of P_t. The values of N, D_0, β_0, l, and μ used in the preceding formulas are based on the values given by Fung (1984) and Zhuang et al. (1983) as modified by Krishnan et al. (1986). Based on the data of Cox (1984), γ was chosen to be 2.0 for the arteries and veins. To calculate C_s for the capillary sheet, Eq. 8.11 was used with h replacing D, and, based on the data of Fung (1984), γ was set equal to 2.3.

With the parameters of the model specified, the steady state R, C_s, and L model distributions are obtained by assigning a value of $\dot{Q}$ and P_v and calculating the intravascular pressures from the vein to the artery. In Figure 8.6, the steady state intralobar vascular pressure distribution for the model is graphed versus the cumulative vascular volume from pulmonary artery to vein for an assigned flow rate of 5.3 ml/s, venous pressure of 1.8 cm H_2O, pleural pressure of -7.6 cm H_2O, and alveolar pressure of 0. The resulting pulmonary artery pressure was approximately 15.6 cm H_2O. The mean capillary pressure, which is generally the intended objective of analysis of the occlusion data, is then 6.4 cm H_2O. Of the total arteriovenous pressure drop, 60% is arterial (mainly the small arteries), 26% is venous (mainly the small veins), and 14% is capillary. In Figure 8.7, the calculated R, L, and C_s distributions for the pressure distribution of Figure 8.6 are also plotted versus

FIGURE 8.7. The vascular resistance (R), inertance (L), and compliance (C_s) of each order divided by the volume of each order plotted versus cumulative vascular volume. (Adapted with permission from ASME/BED *Adv in Bioeng*, Linehan et al., 1988.)

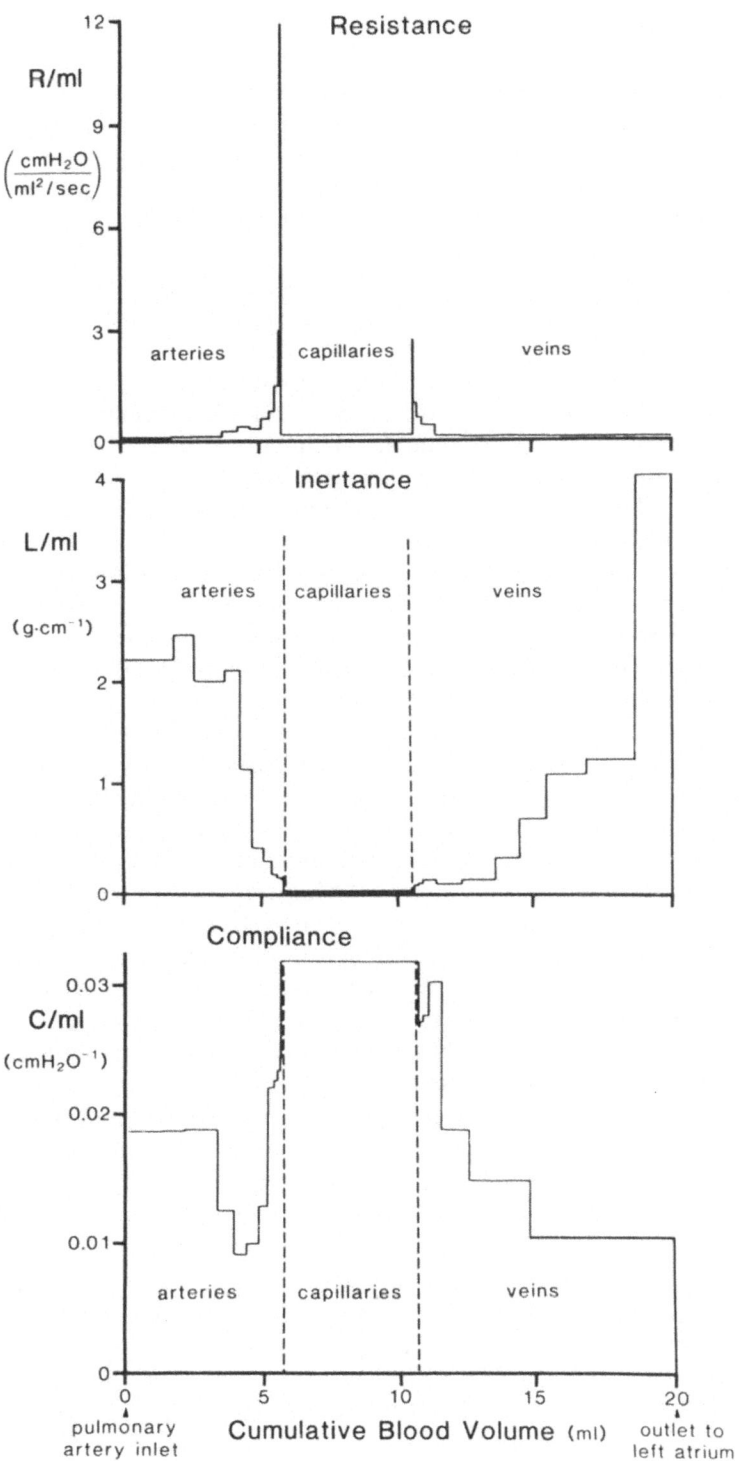

cumulative vascular volume. In this graph, the ordinates are the calculated R, L, and C_s values for each order divided by the calculated volume of that order. Several observations are rather clear from Figures 8.6 and 8.7. Due to their relatively larger diameter, the large arteries and veins have relatively small resistance but most of the fluid inertance. In addition, since the large arteries and veins and the capillary sheet contain the largest volumes, the vascular compliance is concentrated in these regions with very little compliance in the high-resistance small arteries and veins.

The occlusion experiments were simulated by introducing the appropriate step change in either arterial inflow or venous outflow or both and solving the 70 first-order differential equations for the 22 orders of arteries and veins and for the capillary sheet. The differential equations were solved on a DEC 8065 computer using Gear's method for a system of stiff differential equations. The coefficients (R, L, C_c) of the differential equations, which depend on the instantaneous values of vascular volume and pressure, were continuously recalculated within the numerical integration process. At each calculated increment of pressure, volume conservation was required within each order.

In Figures 8.8 through 8.10 selected model simulation results are shown. The arterial pressure (P_a), mean capillary pressure (P_c), and venous pressure (P_v) are graphed for the venous occlusion, double occlusion, and arterial occlusion experiments. In the venous occlusion simulation, wherein the inflow rate is held constant, P_a, P_c, and P_v all increase with time following occlusion as blood is stored in the lung. The first 0.4 s of the simulation is shown in Figure 8.8. In this time frame the effects of the inertance of the blood on the transient pressure curves are manifest. The rapid oscillations in P_v contain frequencies related to the reflections of the pressure and flow waves as they propagate from the point of occlusion, the pulmonary vein. The near absence of oscillations in P_c manifests the impact of dampening within the small veins. We have previously assumed that the arterial and venous pressure curves were free from the effects of fluid inertance about 0.3 s postocclusion (Bronikowski et al., 1984), and the simulation is in agreement with that assumption. After the interval dominated by inertance has passed, there is a period in which a quasi–steady state is achieved. That is, the rate of change in pressure as the vessels fill with blood is relatively constant. During this period $P_a - P_c$ remains larger than $P_c - P_v$ due to the large capillary compliance that accomodates most of the stored blood volume. If the compliance of the veins were very small, the $P_c - P_v$ pressure difference would approach zero. In other words, the $P_c - P_v$ difference is related to the venous compliance as well as the flow through the venous resistance. The results from this simulation of the venous occlusion are consistent with the concept that P_c calculated according to Eq. 8.2 using the quasi–steady state extrapolations is a reasonable representation of the capillary pressure. Note that P_{ai} and P_{vi}, defined as the intersections of the dotted lines in Figure 8.8 with the time of occlusion, can be used in conjuction with Eq. 8.2 to calculate a value of $P_c = 6.1$ cm H_2O, which is in close agreement with the preocclusion mean value of capillary pressure of

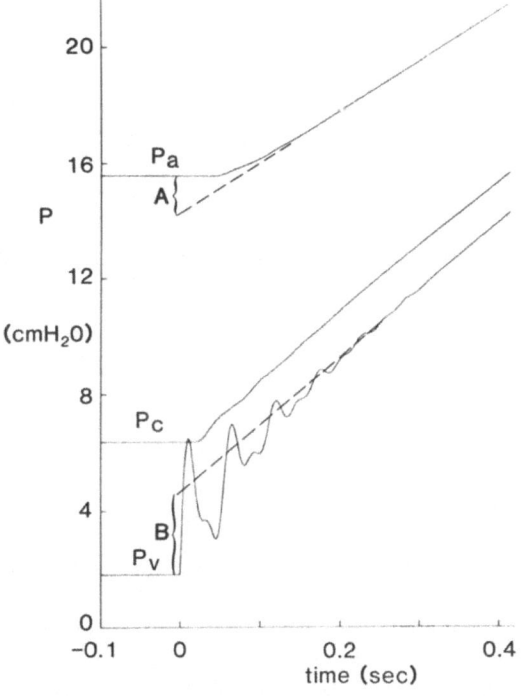

FIGURE 8.8. Simulated venous occlusion. P_a and P_v are the pulmonary artery and vein pressures. P_c is the mean capillary pressure, that is, the pressure at the point midway between the arterial entrance to and venous exit from the capillary sheet. Time is measured from the instant of occlusion. Before occlusion, the flow is constant and the pressures are steady. The dotted lines extrapolated back to $t = 0$ define P_{ai} and P_{vi} (see Fig. 8.1). In this figure, A and B are $P_a(0) - P_{ai}$ and $P_{vi} - P_v(0)$, respectively. The sum $A + B$ is the black box model estimate of $P_c - P_v(0)$ (see Eq. 8.2). The preocclusion value of P_c from the simulation is 6.4 cm H_2O and the estimate of P_c from $A + B$ is 6.1 cm H_2O. (Adapted with permission from ASME/BED *Adv in Bioeng*, Linehan et al., 1988.)

6.4 cm H_2O. When the venous occlusion simulations are carried out over an even longer time interval, additional features of the response can be observed. A more detailed discussion of the interpretable information on the longer time frame in terms of vessel wall viscoelasticity is presented by Krishnan et al. (1986).

In the double-occlusion simulation, the flow both into the pulmonary artery and out of the pulmonary vein is suddenly stopped. The graphs of P_a, P_c, and P_v are shown in Figure 8.9. After a short time, P_a, P_c, and P_v all approach a common asymptotic steady state pressure (P_d). The size of the decrease in P_a and increase in P_v reflect blood volume shifts from arteries to capillaries and from capillaries to veins. Since the arterial and venous compliances are not

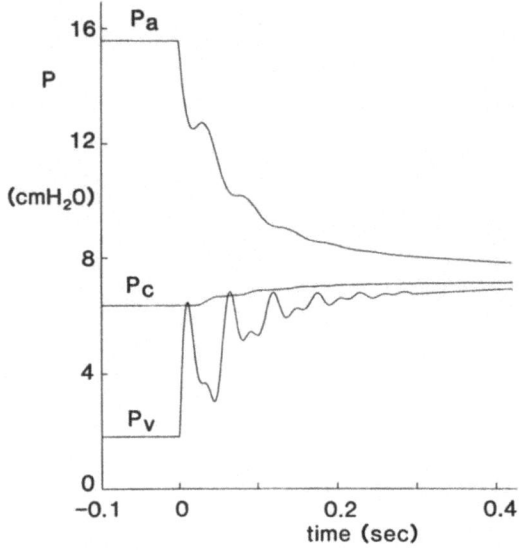

FIGURE 8.9. Simulated double occlusion in the same format as Figure 8.8. $P_d = 7.1$ cm H_2O in comparison to the preocclusion value of $P_c = 6.4$ cm H_2O. (Adapted with permission from ASME/BED *Adv in Bioeng*, Linehan et al., 1988.)

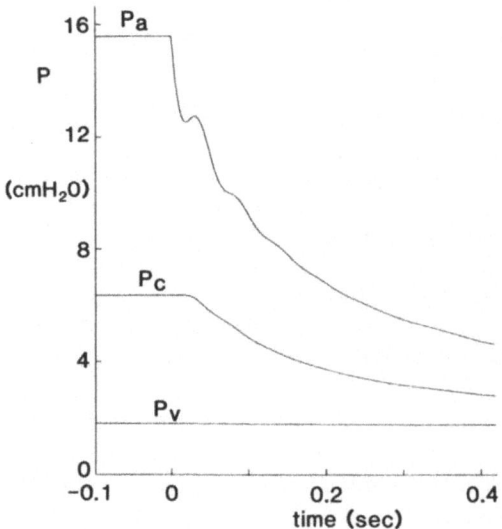

FIGURE 8.10. Simulated arterial occlusion in the same format as Figure 8.8. (Adapted with permission from ASME/BED *Adv in Bioeng*, Linehan et al., 1988.)

markedly different from each other and are relatively small when compared to the capillary compliance, P_a and P_v eventually reach a common asymptotic value P_d, which is also near both the steady state P_c and the P_c calculated using the simulated venous occlusion data from Figure 8.8 and Eq. 8.2. There is also reasonable agreement between the shapes of the simulation results and the data shown in Figure 8.1.

The pressure curve following arterial occlusion also contains information about the distribution of resistance and compliance (Hakim et al., 1982, 1983), and there is considerable interest in approaches for deciphering this information because this maneuver can be carried out in vivo by simply inflating the balloon on a Swan-Ganz catheter (Collee et al., 1987; Cope et al., 1986; D'Orio et al., 1988; Holloway et al., 1983; Siegel and Pearl, 1988). Therefore, attempts have been made to estimate the microvascular pressure using the arterial occlusion data. Figure 8.10 shows the results of a simulated arterial occlusion. Following arterial occlusion, both P_a and P_c decrease as the blood drains from the vein. The overall rate of fall in the arterial pressure is determined mainly by the fraction of the vascular resistance downstream from most of the vascular compliance (see Figure 8.2). However, shortly after arterial occlusion there are also oscillations in the decreasing P_a curve due to the inertance in the large arteries. While the $P_a - P_c$ and $P_c - P_v$ differences after arterial occlusion are related to the relative values of R and C_s among the arteries, capillaries, and veins, it is more difficult to find a simple method, analogous to those used on the venous and double-occlusion pressure curves, to estimate P_c from the measured $P_a(t)$ curve following arterial occlusion. One approach has been to fit the $P_a(t)$ curve using a sum of exponentials. Then certain combinations of the resulting regression coefficients have been used in a model or a correlation equation to estimate P_c. Exactly how best to accomplish this is still a subject of investigation (Collee et al., 1987; Cope et al., 1986; Dawson et al., in press; D'Orio et al., 1988; Holloway et al., 1983; Siegel and Pearl, 1988), which is made complex by the distributed nature of the resistance and compliance in the vascular bed and the fact that there is no steady or quasi–steady state achieved following the arterial occlusion as there is following double occlusion or venous occlusion.

Conclusions

In conclusion, we have shown that a compartmental R, L, C_s model incorporating morphometric and elasticity data can be used to study features of the measureable pressures, P_a and P_v, from occlusion experiments. Such modeling can provide a more detailed perspective underlying the rationale for estimating the microvascular pressure from a black box model analysis of the pressure curves. The dynamic cat lung model provides what appears to be a realistic representation in that it includes considerable detail of the vascular bed relevant to the occlusion responses. The results of simulations with this model appear to be consistent with data interpretations of venous and double-occlusion data suggested by less detailed (black box) models. The arterial occlusion maneuver, which is clinically practical, offers interesting potential for clinical physiological interpretation. Unfortunately, even with high-resolution data analysis, the interpretation of the transient data immediately after arterial occlusion in terms of the arterial resistance may be confounded by the dominating influence of fluid inertance and vascular compliance. At later times, the measurable P_a decay curve will be multiexponential due to the continuous distribution of resistance and compliance among the series-connected orders of vessels. This may introduce additional complexity in the interpretation of the longer time data to obtain useful estimates of P_c than in the venous and double-occlusion experiments. However, the use of a morphometrically based model, like the cat lung dynamic model presented, may be helpful in establishing a basis for estimating P_c from the arterial occlusion data as well.

Acknowledgment.   This study was supported by National Heart, Lung and Blood Institute grant HL-19298 and the Research Service of the Veterans Administration.

References

Bronikowski TA, Dawson CA, Linehan JH (1985) Limits on continuous distribution of pulmonary vascular resistance versus compliance from outflow occlusion. *Microvasc Res* 30:306–313.

Bronikowski TA, Linehan JH, Dawson CA (1984) A model of the vascular resistance and compliance distribution in a lung lobe. *Microvasc Res* 28:289–310.

Bshouty Z, Ali J, Younes M (1987) Arterial occlusion versus isofiltration pulmonary capillary pressures during very high flow. *J Appl Physiol* 62:1174–1178.

Collee GG, Lynch KE, Hill RD, Zapol WM (1987) Bedside measurement of pulmonary capillary pressure in patients with acute respiratory failure. *Anesthesiology* 66:614–620.

Cope DK, Allison RC, Parmentier JL, Miller JN, Taylor AE (1986) Measurement of effective pulmonary capillary pressure using the pressure profile after pulmonary artery occlusion. *Crit Care Med* 14:16–22.

Cox RH (1984) Viscoelastic properties of canine pulmonary arteries. *Am J Physiol* 246:90–96.

Dawson CA, Bronikowski TA, Linehan JH, Haworth ST, Rickaby DA (in press) On estimation of pulmonary capillary pressure from arterial occlusion. *Am Rev Resp Dis.*

Dawson CA, Linehan JH, Rickaby DA (1982) Pulmonary microcirculation hemodynamics. *Ann NY Acad Sci* 384:80–106.

D'Orio V, Halleux J, Rodriguez LM, Wahlen C, Marcelle R (1988) Effects of *Escherichia coli* endotoxin on pulmonary vascular resistance in intact dogs. *Crit Care Med* 14:802–806.

Fung YC (1984) *Biodynamics: Circulation.* Springer-Verlag, New York.

Gaar KA, Taylor AE, Owens LJ, Guyton AC (1967) Pulmonary capillary pressure and filtration coefficient in the isolated perfused lung. *Am J Physiol* 213:910–914.

Hakim TS, Dawson CA, Linehan JH (1979) Hemodynamic responses of a dog lung lobe to lobar venous occlusion. *J Appl Physiol* 47:142–152.

Hakim TS, Michel RP, Chang HK (1982) Partitioning of pulmonary vascular resistance in dogs by arterial and venous occlusion. *J Appl Physiol* 52:710–715.

Hakim TS, Michel RP, Minami H, Chang HK (1983) Sites of pulmonary hypoxic vasoconstriction studied with arterial and venous occlusion. *J Appl Physiol* 54:1298–1302.

Holloway H, Perry M, Downey J, Parker J, Taylor A (1983) Estimation of effective pulmonary capillary pressure in intact lungs. *J Appl Physiol* 54:846–851.

Krishnan A, Linehan JH, Rickaby DA, Dawson CA (1986) Cat lung hemodynamics: Comparison of experimental results and model predictions. *J Appl Physiol* 61:2023–2034.

Linehan JH, Dawson CA (1983) A three compartment model of the pulmonary vasculature: Effects of vasoconstriction. *J Appl Physiol* 55:923–928.

Linehan JH, Dawson CA, Rickaby DA (1982) Distribution of vascular resistance and compliance in a dog lung lobe. *J Appl Physiol* 53:158–168.

Linehan JH, deMora F, Bronikowski TA, Dawson CA (1988) Hemodynamic modelling of vascular occlusion experiments in cat lung. ASME. BED *Adv in Bioeng* 8:139–142.

Milnor WR (1982) *Hemodynamics*. Williams & Wilkins, Baltimore.

Parker JC, Kvietys PR, Ryan KP, Taylor AE (1983) Comparison of isogravimetric and venous occlusion capillary pressures in isolated dog lungs. *J Appl Physiol* 55:964–968.

Rippe B, Parker JC, Townsley MI, Mortillaro NA, Taylor AE (1987) Segmental vascular resistances and compliances in dog lung. *J Appl Physiol* 62:1206–1215.

Rock P, Patterson GA, Permutt S, Sylvester JT (1985) Nature and distribution of vascular resistances in hypoxic pig lungs. *J Appl Physiol* 59:1891–1901.

Siegel LC, Pearl RG (1988) Measurement of the longitudinal distribution of pulmonary vascular resistance from pulmonary artery occlusion pressure profiles. *Anesthesiology* 68:305–307.

Zhuang FY, Fung YC, Yen RT (1983) Analysis of blood flow in cat's lung with detailed anatomical and elasticity data. *J Appl Physiol* 55:1341–1348.

9
Sites of Vasoactivity in the Pulmonary Circulation Evaluated Using a Low-Viscosity Bolus Method

Christopher A. Dawson, Thomas A. Bronikowski, John H. Linehan, and David A. Rickaby

Introduction

Because of the importance of pulmonary capillary pressure in the fluid balance of the lungs and the propensity for various pulmonary vasomotor stimuli to cause constriction of pulmonary veins (Dawson, 1984), there has been considerable interest in methods for determining pulmonary capillary pressure and the arteriovenous sites of pulmonary vasoconstriction. A number of approaches have been used, and each approach has had advantages and disadvantages (Agostoni and Piiper, 1962; Bhattacharya and Staub, 1980; Brody et al., 1968; Bronikowski et al., 1985; Dawson et al., 1988; Gaar et al., 1967; Gable and Drake, 1978; Kadowitz et al., 1975; McDonald and Butler, 1967; Michel et al., 1984; Nagasaka et al., 1984; Piiper, 1970; Zhuang et al., 1983). Like several other methods, the low-viscosity bolus method has been an experimental method used in studies of pump-perfused lungs. In such studies it has the potential for providing some unique insights into the influence of vasomotion on the longitudinal distribution of pulmonary vascular resistance and intravascular pressure from pulmonary artery to pulmonary veins. The method, originally introduced by Piiper (1970), has been modified by Brody et al. (1968) and Grimm et al. (1977) and more recently by us (Dawson et al., 1988) in an attempt to improve resolution to take advantage of this potential.

One aspect of this method that has been problematic in the past is that the algebraic system of equations that needs to be solved to obtain the distribution of resistance with respect to vascular volume tends to be ill conditioned. Previously, the methods used to obtain a solution have resulted in the overall method having less resolution (i.e., producing distributions with less structure) than the data themselves appeared to suggest. Recently, we have utilized regularization methods to stabilize the solution and improve the resolution (Dawson et al., 1988; Varah, 1979; Wagner, 1982). The results have been encouraging to us and the present chapter is an attempt to describe the methods presently in use and to show some applications.

Hemodynamic Model

The basic approach involves the measurement of the change in the arterio-venous pressure gradient following the introduction of a bolus of saline, plasma, or blood diluted with saline or plasma into the blood flowing into the pulmonary artery of a pump-perfused lung (Dawson et al., 1988). The pump provides a constant flow so that the time-varying arteriovenous pressure difference reflects the decrease in pulmonary vascular resistance caused by the bolus. The magnitude of this decrease will follow a time course that depends on the longitudinal location of the bolus within the vascular bed at a given time and the preinjection resistance at that location. Thus, the decrease in the arteriovenous pressure difference will tend to be largest when the bolus is located in the regions of highest preinjection resistance. Figure 9.1 shows how the shape of the pressure curve can change in response to vasoconstriction. In this example, the change in shape resulted from the infusion of histamine, which is predominantly a venous constrictor in this preparation. The objective, then, is to determine the longitudinal distributions of vascular resistance with respect to vascular volume from the arterial inlet to venous outlet from these pressure versus time curves that result from the passage of the bolus.

Representing the longitudinal resistance distribution as a finite sum of individual serial resistances, the arteriovenous pressure gradient is related to flow and the time-varying resistance distribution due to passage of the bolus by

$$P_a(t) - P_v(t) = \dot{Q} \sum_{k=1}^{N} R_k(t) \tag{9.1}$$

where P_a is arterial pressure, P_v is venous pressure, $\dot{Q}$ is the steady flow through the lung, and R_k is the resistance of the kth serial segment. R_k changes with time as the viscosity, μ, within segment k changes with time during passage of the bolus. Thus, the change in resistance, $\Delta R_k(t)$, due to the bolus passage is

$$\Delta R_k(t) = \alpha_k \Delta\mu_k(t), \tag{9.2}$$

where α_k is the segmental geometric factor and $\Delta\mu_k(t)$ is the change in viscosity due to the mixture of bolus and blood in the segment. Thus, the change in the arteriovenous pressure gradient, $\Delta(P_a - P_v)(t)$, from its preinjection value can be expressed as

$$\Delta(P_a - P_v)(t) = \dot{Q} \sum_{k=1}^{N} \alpha_k \Delta\mu_k(t) \tag{9.3}$$

For Poiseuille flow in a cylindrical tube α_k would be determined by the length (L) and diameter (D) of the kth segment according to $128L/\pi D^4$. However, knowledge of the functional form relating geometry and resistance in the vascular bed is not necessary for this method as long as the resistance is linearly proportional to viscosity and the α_k values do not vary with time.

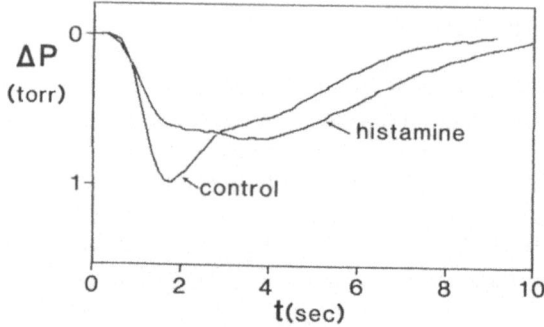

FIGURE 9.1. The change in the arteriovenous pressure difference, $\Delta(P_a - P_v)$ referred to as ΔP on this graph, following the introduction of a 2.5-ml bolus of saline-diluted blood through an isolated dog lung lobe perfused with blood at a flow rate of about 6 ml/sec. The infusion of histamine caused a marked change in the shape of the curve due to a shift in the site of major resistance toward the venous end of the vascular bed. The hematocrits of the two boluses were adjusted so that even though the total vascular resistance was higher during histamine infusion the areas above the ΔP curves were similar.

Given the measured values of $\Delta(P_a - P_v)(t)$ and $\dot{Q}$, the α_k values could be calculated from a system of N simultaneous algebraic equations if the $\Delta\mu_k(t)$ values were known.

Method for Determining Segmental Viscosity

Our approach to finding the $\Delta\mu_k(t)$ values is first to measure the electrical conductivity of the blood at the inlet to the pulmonary artery and at the exit from the pulmonary veins. Due to the relatively poor electrical conductivity of the blood cells, the electrical conductivity of the blood is inversely proportional to the hematocrit (Okada and Schwan, 1960). If the hematocrit is not too high, a reasonably linear relationship between hematocrit and viscosity is obtained. Thus, the output of the conductivity bridge is nearly linearly proportional to viscosity (Grimm et al., 1977). Figure 9.2 shows an example of the pressure and arterial and venous viscosity data collected from an isolated dog lung lobe, wherein the changes in arterial and venous viscosities were determined from the changes in arterial and venous conductivities produced by the bolus.

With the electrical conductivity curves used to estimated the changes in viscosity at the inlet, $\Delta\mu_a(t)$, and outlet, $\Delta\mu_v(t)$, the problem becomes one of using this data to estimate the $\Delta\mu_k(t)$ values within the organ. This is accomplished by assuming that the relationship between the inflow and outflow viscosity curves can be expressed by the convolution integral

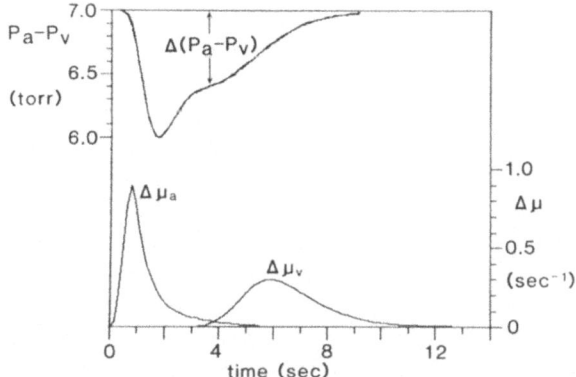

FIGURE 9.2. An example of the changes in the arteriovenous pressure difference, $\Delta(P_a - P_v)$, and in the arterial and venous viscosities with time as a bolus of saline passed through a dog lung lobe. The change in viscosity at the arterial inlet, $\Delta\mu_a$, and at the venous outlet, $\Delta\mu_v$, were estimated from measurements of the electrical conductivity of the blood. These viscosity curves are normalized to unit area on this graph. (From Dawson et al., 1989; reprinted by courtesy of Marcel Dekker, Inc.)

$$\Delta\mu_v(t) = \int_0^t \Delta\mu_a(\lambda) \, h(t - \lambda) \, d\lambda \qquad (9.4)$$

where $h(t)$ is the organ transport function, that is, the frequency function of bolus transit times within the lung. First, $h(t)$ is obtained by numerical deconvolution. This is accomplished by replacing Eq. 9.4 by an equivalent system of linear equations. With both the $\Delta\mu_a(t)$ and $\Delta\mu_v(t)$ digitized with the same time increment (Δt), let $t_j = j\Delta t$ for j ranging from 1 to n, where t_1 is the first time at which $\Delta\mu_a(t) > 0$, and t_n is the last time at which $\Delta\mu_v(t) > 0$. Then using the notation $u_j = \Delta\mu_a(t_j) \, \Delta t$, $d_j = \Delta\mu_v(t_j)$, and $h_j = h(t_j)$, evaluating Eq. 9.4 at times $t_2, t_3 \ldots t_n$ and representation of the convolution integral by a sum results in the following system

$$
\begin{aligned}
u_1 h_1 &= d_2 \\
u_2 h_1 + u_1 h_2 &= d_3 \\
\vdots \qquad\qquad &\quad \vdots \\
u_{n-1} h_1 + u_{n-2} h_2 + \cdots + u_1 h_{n-1} &= d_n
\end{aligned}
\qquad (9.5)
$$

In matrix notation system 9.5 becomes

$$U\mathbf{h} = \mathbf{d} \qquad (9.5a)$$

where $\mathbf{h} = (h_1, h_2, \ldots, h_{n-1})^T$, $\mathbf{d} = (d_2, d_3, \ldots, d_n)^T$, and U is the $(n-1) \times (n-1)$ matrix with entries $U_{ij} = u_{i-j+1}$ if $j \leq i$, or 0 if $j > i$.

System 9.5 has a unique solution $\mathbf{h}$ that can be obtained recursively, but, in general, such solutions become physically unrealistic in the face of even the slightest amount of noise in the experimental data. When Δt is small, n is large and the recursive solution of system 9.5 tends to exhibit oscillations with large positive and negative values. Therefore, we apply a regularization method to obtain estimates of $h(t)$. This method is derived from the minimization of a sum of two terms. The first is the sum of the squared differences of the left and right sides of the equations in system 9.5. When this first term is small then Eq. 9.4 and system 9.5 are closely satisfied. The second term is of the form $Z \sum h_j^2$ where $Z > 0$. When this term is small, the oscillations of the direct solution of system 9.5 tend to disappear. Thus, the simultaneous minimization of the sum of these two sums yields, for a fixed Z, a better approximation to $h(t)$ than the solution of system 9.5. The Z-dependent vector $\mathbf{h}$ that minimizes this sum of terms can be shown to satisfy the system:

$$(U^T U + ZI)\mathbf{h} = U^T \mathbf{d} \tag{9.6}$$

where I is the identity matrix. As Z approaches 0, the solutions of system 9.6 approach the highly oscillating solution of system 9.5; as Z becomes large, the solutions tend to the zero vector, which is not appropriate as a transport function. Thus, somewhere between the extremes of very small and very large values of Z there is a Z that is optimal with respect to two goals: (1) that the corresponding $\mathbf{h}$ be unimodular, and (2) that $\mathbf{h}$ reproduce the measured $\Delta\mu_v(t)$ data well when convolved with $\Delta\mu_a(t)$. Finding the optimal Z value depends on quantifying the degree to which these two goals are satisfied.

We quantify the oscillatory behavior using the quantity F_1, where

$$F_1 = \sum_{j=2}^{n-1} |h_j - h_{j-1}| - 2h_{max}$$

The sum in F_1 is the total variation of the estimated $h(t)$ as represented by the vector $\mathbf{h}$. The second term is the total variation of a function that rises monotonically from 0 to a peak value, h_{max}, then decreases back to 0. F_1 is therefore the excess of total variation over that of a unimodular function. A large value of F_1 means that the current $h(t)$ estimate oscillates. A small value of F_1 implies few oscillations of small amplitude. F_1 tends to be larger when Z is small and approaches 0 when Z is large.

The appropriateness of $h(t)$ as a transport function is quantified by the coefficient of variation, denoted by F_2, between the measured $\Delta\mu_v(t)$ data and the convolution of $\Delta\mu_a(t)$ with the estimate of $h(t)$. The behavior of F_2 as Z varies is opposite that of F_1; F_2 is small when Z is small and large when Z is large. Consequently, the sum $F_1 + F_2$ is large when Z is either too small or too large. Thus, the optimal Z value is found by solving system 9.6 over a range of Z values until the sum $F_1 + F_2$ is minimized. Figure 9.3 shows an example of a transport function obtained in this manner.

Having obtained the organ transport function by solving system 9.6 in this fashion, we next specify the number, N, of equal serial vascular volume segments. We have generally set $N = 20$. We then define the transport func-

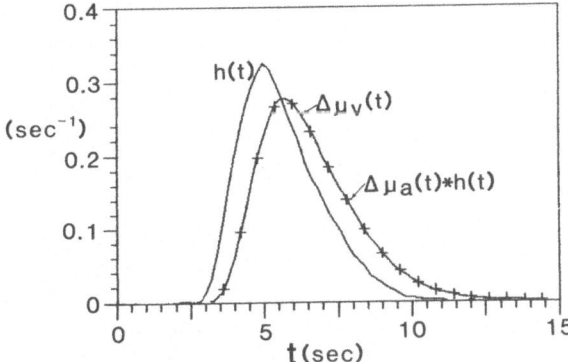

FIGURE 9.3. An example of a transport function, $h(t)$, obtained by numerical deconvolution of the $\Delta\mu_a(t)$ and $\Delta\mu_v(t)$ curves. The measured $\Delta\mu_v(t)$ curve is also shown along with the convolution of the $\Delta\mu_a(t)$ curve with $h(t)$, $\Delta\mu_a(t) * h(t)$. Comparison of the latter two curves is indicative that the calculated $h(t)$ is a reasonable estimate of the organ transport function. (From Dawson et al., 1989; reprinted by courtesy of Marcel Dekker, Inc.)

tion, $h_k(t)$, from the inlet of the lobe to the end of the kth volume segment in terms of the lung transport function as follows:

$$h_k(t) = (N/k)h\left((N/k)t\right) \tag{9.7}$$

These intermediate transport functions have the same relative dispersion as the whole lobe, and mean transit times and variances appropriate for the volume extending from the lobe inlet through the kth volume segment. Thus, if $\bar{t}$ and σ^2 are the mean transit time and variance of $h(t)$, respectively, then $\bar{t}_k$ and $\sigma^2{}_k$, the mean transit time and variance of $h_k(t)$, are given by $\bar{t}_k = (k/N)\bar{t}$ and $\sigma^2{}_k = (k/N)^2\sigma^2$, respectively. With these transport functions, the $\Delta\mu_k(t)$ at the various volume segments is calculated as a convolution of $\Delta\mu_a(t)$ and $h_k(t)$:

$$\Delta\mu_k(t) = \int_0^t \Delta\mu_a(\lambda)\, h_k(t - \lambda)\, d\lambda \tag{9.8}$$

Figure 9.4 is an example showing the measured $\Delta\mu_a$ and $\Delta\mu_v$ curves from a dog lung lobe, and four of the 19 calculated $\Delta\mu_k(t)$ curves obtained using this method.

Method for Determining Segmental Resistance Distribution

Once the $\Delta\mu_k(t)$ estimates are available, we can proceed to calculate the preinjection resistances, $R_k(0)$. With the measured $\Delta(P_a - P_v)(t)$ data digitized at the M equally spaced times $t_1, t_2, \ldots, t_M$, where $t_i = i\Delta t$, substituting into

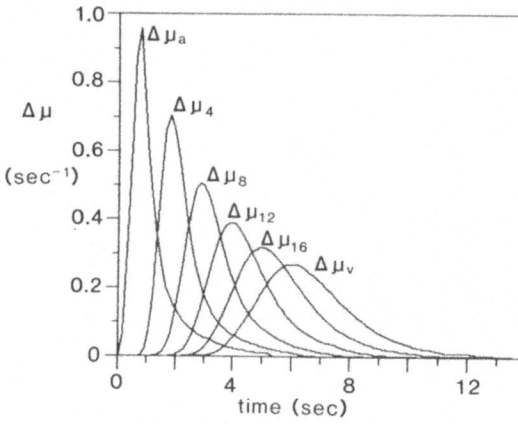

FIGURE 9.4. An example of the measured $\Delta\mu_a(t)$ and $\Delta\mu_v(t)$ data and four of the $\Delta\mu_k(t)$ curves calculated using the procedure described in the text. The $\Delta\mu_k(t)$ curves are those that, according to the model, would be measured at the outlets of volume segments 4, 8, 12, and 16 if it were possible to sample appropriately within the organ. (From Dawson et al., 1989; reprinted by courtesy of Marcel Dekker, Inc.)

Eq. 9.3 we obtain a linear system of M equations in the N unknown α_k values. Eq. 9.3 can be expressed as

$$p_i = \sum_{k=1}^{N} g_{ik} r_k \tag{9.9}$$

where

$$p_i = \Delta(P_a - P_v)(t_i) \Big/ \int_0^\infty \Delta(P_a - P_v)(t)\, dt$$

$$g_{ik} = \Delta\mu_k(t_i) \Big/ \int_0^\infty \Delta\mu(t)\, dt$$

$$r_k = \alpha_k \Big/ \sum_{k=1}^{N} \alpha_k$$

Introducing the vectors $\mathbf{p} = (p_1, \ldots, p_M)^T$, and $\mathbf{r} = (r_1, \ldots, r_n)^T$ and matrix $G = (g_{ik})$, we may rewrite system 9.9 in matrix notation as

$$\mathbf{p} = G\mathbf{r} \tag{9.9a}$$

Despite the overdetermined nature of system 9.9, the presence of even a small degree of noise in the experimental conductance and pressure data causes even the standard least squares solution of this system to yield physically meaningless results. In previous studies this problem was solved either by choosing a large Δt (Brody et al., 1968) or by assuming a functional form for $\mathbf{r}$ (Grimm et al., 1977). However, both of these approaches produce relatively poor resolution so that the details of the structure of the longitudinal resistance distribution are lost. Therefore, we again utilize the regularization method described above in reference to the deconvolution to estimate $\mathbf{r}$. In the present case, system 9.10, which is analogous to system 9.6, is solved over a range of Z values:

$$(G^T G + ZI)\mathbf{r} = G^T \mathbf{p} \tag{9.10}$$

Since we can make no a priori assumptions about the shape of the $\mathbf{r}$ distribution other than that r_k is positive, we cannot use F_1 in determining the optimal solution. Instead only the coefficient of variation, F_2, between measured ΔP data and values of ΔP calculated from system 9.9 is used. Because of nonnegativity constraints, any negative r_k entries in the solution are set equal to 0 before F_2 is calculated. Thus, when Z is small some negative r_k values may occur. When these are eliminated, F_2 tends to be large. As Z is increased, the negative r_k values disappear from the solution, and F_2 decreases. When Z becomes too large F_2 begins to increase again. To find the optimal solution we increase Z successively until the negative values of r_k disappear, F_2 is near its minimum, and there is a smooth distribution of r_k. Since r_k is proportional to $R_k(0)$,

$$R_k(0) = r_k \left(\sum_{k=1}^{N} R_k(0) \middle/ \sum_{k=1}^{N} r_k \right) \tag{9.11}$$

where

$$\sum_{k=1}^{N} R_k(0) = (P_a - P_v)/\dot{Q}$$

prior to the injection.

To determine the values of $\dot{Q}_k$, the total blood volume of the lung lobe ($\dot{Q}_L$) is calculated as the mean transit time of $h(t)$ times the flow rate, $\dot{Q}$. Then $\dot{Q}_k = \dot{Q}_L/N$. Figure 9.5 depicts the segmental vascular resistance per milliliter

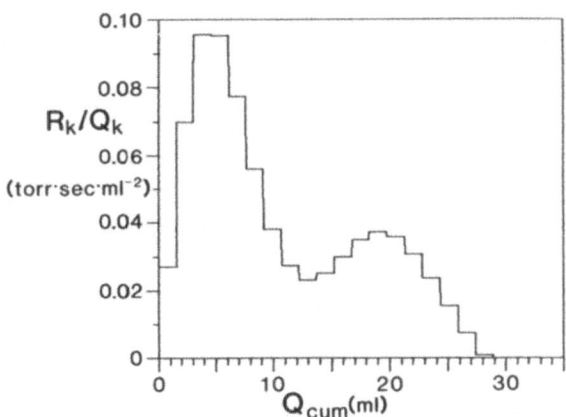

FIGURE 9.5. An example of a resistance distribution obtained from a dog lung lobe in the normal vasodilated state with a lung volume near functional residual capacity. The vascular resistance is in terms of the segmental resistance per unit segmental volume. The $\dot{Q}_{cum}$ is the cumulative vascular volume starting with the arterial inlet at $\dot{Q}_{cum} = 0$ and extending to the venous outlet, which in this case is about 29 ml. The bimodal distribution is typical for a dog lung lobe under these conditions. (From Dawson et al., 1989; reprinted by courtesy of Marcel Dekker, Inc.)

of blood in each of the several volume segments as a function of the cumulative blood volume, Q_{cum}, from the arterial inlet at 0 to the venous outlet at $\dot{Q}_L$. The value of $\dot{Q}_{cum}$ is then $\dot{Q}$ times the mean transit time of $h_k(t)$, that is, the volume from arterial inlet to the end of volume segment k.

Illustrative Results

The resistance distribution shown in Figure 9.5 is typical of that obtained from an isolated dog lung lobe at a lung volume near functional residual capacity (Dawson et al., 1988). We have interpreted the bimodal distribution obtained under these conditions as being indicative of relatively high-resistance regions upstream and downstream from the capillary bed. Due to the heterogeneity of parallel pathways traveled by the bolus as it passes through the lungs, the kth volume segment contains a mix of different kinds of vessels. Near the entrance at the arterial inlet, where $\dot{Q}_{cum}/\dot{Q}_L$ is small, $\dot{Q}_k$ contains mainly large arteries. As k increases $\dot{Q}_k$ becomes a mix of large and small arteries, then small arteries and capillaries, then capillaries and small veins, and so on. Thus, it is not possible to designate the location of the capillary end of the arterial tree or the beginning of the veins on the $\dot{Q}_{cum}/\dot{Q}_L$ axis. In addition, the ana-tomical mix in a particular $\dot{Q}_k$ will not necessarily remain constant when the resistance distribution is changed by changes in the vascular geometry due to mechanical manipulations or vasomotion. On the other hand, the relative changes in arterial and venous resistance do appear to be revealed quite well.

Figures 9.6 through 9.9 show the results obtained from one dog lung lobe

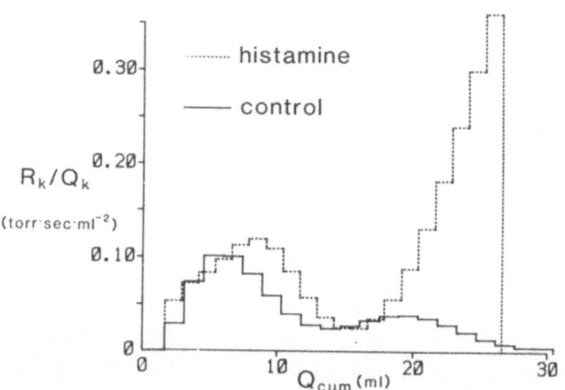

FIGURE 9.6. The resistance distribution obtained during histamine infusion compared with that obtained during normal control conditions. Histamine caused primarily downstream constricting, presumably mainly venous constriction. Same format as for Figure 9.5.

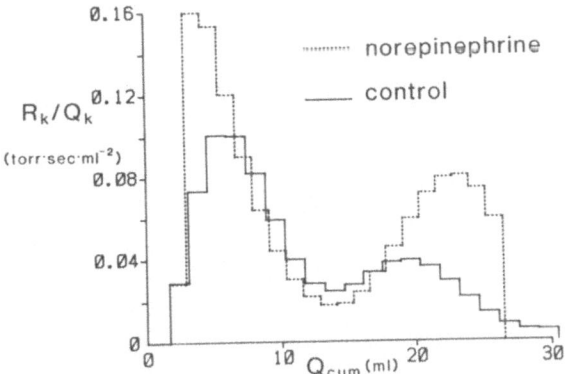

FIGURE 9.7. The resistance distribution obtained during norepinephrine infusion compared with that obtained during normal control conditions. Norepinephrine caused both upstream and downstream constriction, presumably both arterial and venous constriction. Some format as for Figure 9.5. (From Dawson et al., 1989; reprinted by courtesy of Marcel Dekker, Inc.)

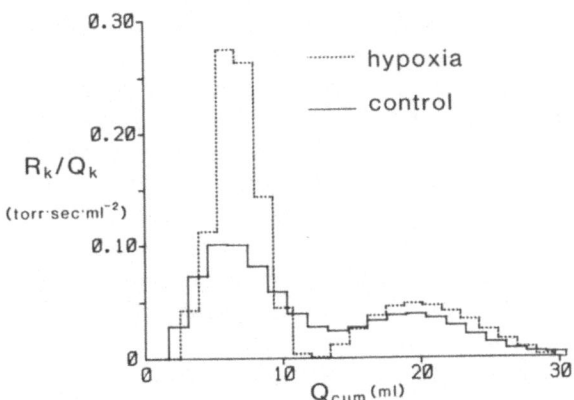

FIGURE 9.8. The resistance distribution obtained during hypoxia ($P_AO_2 = 35$ torr) compared with that obtained during normal control conditions ($P_AO_2 = 111$ torr). Hypoxia caused primarily upstream constriction, presumably arterial constriction. Same format as for Figure 9.5.

with four different vasoconstrictor stimuli each having different relative potencies at different longitudinal sites along the vascular bed. In this experiment the drug infusion rates and Po_2 were adjusted to increase the total vascular resistance by about 50%. The fact that there is a centrally located nadir in the bimodal distribution despite the differences in the longitudinal resistance distributions suggests that there is a region between the muscular arteries and

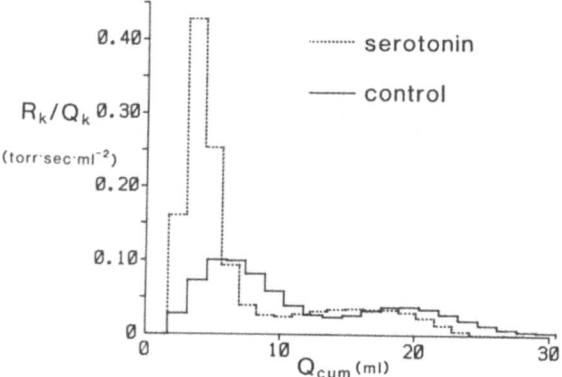

FIGURE 9.9. The resistance distribution obtained during serotonin infusion compared with that obtained during normal control conditions. Serotonin caused primarily upstream constriction, presumably arterial constriction. Same format as for Figure 9.5. (From Dawson et al., 1989; reprinted by courtesy of Marcel Dekker, Inc.)

veins that does not respond to these stimuli, namely, the capillary bed. Thus, it would appear that the pressure drop downstream from this nadir is probably close to the pressure drop downstream from the mean capillary pressure. Figure 9.10 shows how the intravascular pressure distribution was affected by constriction with histamine or serotonin. According to the concept proposed above, the centrally located inflection points of the pressure distributions in Figure 9.10, which are at the nadirs on the resistance versus cumulative volume graphs, should be near the mean capillary pressure. In the case of serotonin infusion the capillary pressure was not changed much from the control condition, whereas with histamine infusion it was substantially increased. It can be further noted that the location of the capillary pressure in terms of the cumulative volume was altered by the vasoconstrictors. This is consistent with the concept that, with the predominantly venous constrictor histamine, there was some distention of capacitance vessels upstream, whereas with the arterial constrictor serotonin, the upstream volume decreased due to displacement of volume out of the constricted vessels. The resistance distributions shown in Figure 9.7 suggest that norepinephrine constricted both arteries and veins. A comparison between Figures 9.8 and 9.9 suggests that while both hypoxia and serotonin increased the arterial resistance, serotonin may have decreased the arterial compliance as well. This is based on the observation that during serotonin infusion the segments having the highest resistance were at smaller values of $\dot{Q}_{cum}$ than during either the control or hypoxic conditions. Thus it appears that this method can distinguish between arterial and venous sites of vasomotion and provide some additional details about the effects of vasoconstrictor stimuli on the pulmonary vascular mechanics.

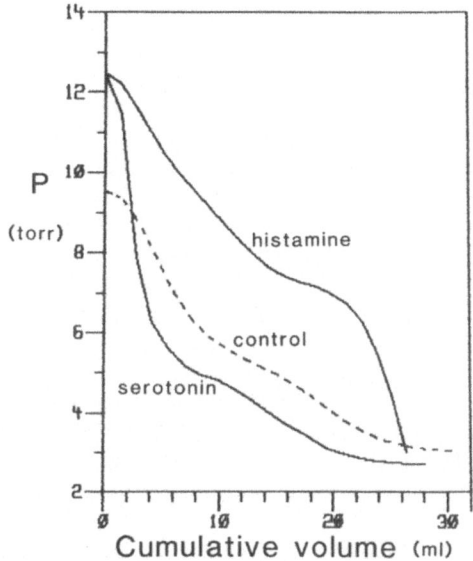

FIGURE 9.10. The intravascular pressure, P, from the arterial inlet (cumulative volume = 0) to the venous outlet (maximum cumulative volume) under normal control conditions and during histamine and serotonin infusion. Assuming that the inflection point near the middle of these pressure distributions is near the mean capillary pressure (i.e., the nadir in the resistance distributions in Figure 9.6 and 9.9), one can see how the vasoconstriction influenced both the magnitude of the capillary pressure and the location of the capillary pressure along the cumulative volume distribution.

Acknowledgment. This study was supported by National Heart, Lung and Blood Institute grant HL-19298 and the Research Service of the Veterans Administration.

References

Agostoni E, Piiper J (1962) Capillary pressure and distribution of vascular resistance in isolated lung. *Am J Physiol* 202:1033–1036,

Bhattacharya J, Staub NC (1980) Direct measurement of microvascular pressures in isolated perfused dog lung. *Science* 210:327–328.

Brody JS, Stemmler EJ, DuBois AB (1968) Longitudinal distribution of vascular resistance in pulmonary arteries, capillaries and veins. *J Clin Invest* 47:783–799.

Bronikowsk TA, Dawson CA, Linehan JH (1985) Limits on continuous distribution of pulmonary vascular resistance versus compliance from outflow occlusion. *Microvasc Res* 30:306–313.

Dawson CA (1984) Role of pulmonary vasomotion in physiology of the lung. *Physiol Rev* 64:544–616.

Dawson CA, Bronikowski TA, Linehan JH, Rickaby DA (1988) Distributions of vascular pressure and resistance in the lung. *J Appl Physiol* 64:274–284.

Dawson CA, Linehan JH, Bronikowski TA (1989) Pressure and flow in the pulmonary vascular bed. In Weir EK, Reeves JT (eds) *Pulmonary Vascular Physiology and Pathophysiology*. Marcel Dekker, New York, pp 51–105.

Gaar KA, Taylor AE, Owens LJ, Guyton AC (1967) Pulmonary capillary pressure and filtration coefficient in the isolated perfused lung. *Am J Physiol* 213:910–914.

Gable JC, Drake RE (1978) Pulmonary capillary pressure in intact dog lungs. *Am J Physiol* 235:H569–H573.

Grimm DJ, Linehan JH, Dawson CA (1977) Longitudinal distribution of vascular resistance in the lung. *J Appl Physiol* 43:1093–1101.

Kadowitz PJ, Joiner PD, Hyman AL (1975) Influence of sympathetic stimulation and vasoactive substances on the canine pulmonary veins. *J Clin Invest* 56:354–365.

McDonald IG, Butler J (1967) Distribution of vascular resistance in the isolated perfused dog lung. *J Appl Physiol* 23:463–474.

Michel RP, Hakim TS, Chang HK (1984) Pulmonary arterial and venous pressures measured with small catheters in dogs. *J Appl Physiol* 57:309–314.

Nagasaka Y, Bhattacharya J, Nanjo S, Gropper MA, Staub NC (1984) Micropuncture measurement of lung microvascular pressure profile during hypoxia in cats. *Circ Res* 54:90–95.

Okada RH, Schwan HP (1960) An electrical method to determine hematocrits. *IRE Trans Med Electron* 7:188–192.

Piiper J (1970) Attempts to determine volume, compliance, and resistance to flow of pulmonary vascular compartments. *Prog Resp Res* 5:40–52.

Varah JM (1979) A practical examination of some numerical methods for linear discrete ill-posed problems. *SIAM Rev* 21:100–111.

Wagner PD (1982) Calculating the distribution of ventilation-perfusion ratios from inert gas elimination data. *Fed Proc* 41:136–139.

Zhuang FY, Fung YC, Yen RT (1983) Analysis of blood flow in cat's lung with detailed anatomical and elasticity data. *J Appl Physiol* 55:1341–1348.

10
Volumetric Changes of Pulmonary Capillaries as Assessed by a Density Method

JEN-SHIH LEE and LIAN-PIN LEE

Introduction

We have observed that an intermittent positive pressure ventilation of the dog produces a cyclic variation of the same frequency in the density of arterial blood (Lee and Lee, 1986). To provide the basics leading to this observation, we first describe the density measuring system, the animal experimentation, and the use of the density difference as a concentration of a "density" indicator to quantify the cardiac output and the mean transit time.

Based on the presence of the Fahraeus effect in capillary blood flow, we have demonstrated theoretically that the ventilatory-induced changes in the blood volume of the pulmonary capillaries (V_c) could produce a density variation in the blood leaving the capillaries. If we regard this variation as one introduced to a site C within the capillaries, this variation is carried by the blood and dispersed to form the density variation in the aortic blood (Lee, 1988). An estimation of the dispersion from site C to the sampling site is made by appropriately fractionalizing the dilution curve of the central circulation, which consists of the two heart chambers and the pulmonary vasculature. Through the use of a Fourier series to represent the sampled density variation and a Fourier transform to perform the dispersion analysis, we determine the variation in the blood leaving the capillaries and hence the volumetric deformation of the pulmonary capillaries.

In essence, *this density method and analytic procedure open a new way to assess the change in the blood volume of the pulmonary capillaries.* In comparison, the method using carbon monoxide uptake (Roughton and Forster, 1957), may not be able to recognize the small volumetric change in the capillaries as induced by the ventilation. The measurement of the transit time of oxygen through the venous and arterial networks of the lung could yield an estimate of the mean transit time through the pulmonary capillaries and hence their volume (Piiper, 1970). However, the incorporation of this transit time measurement into in vivo experimentation is difficult.

Experiments with rabbits subject to intermittent positive-pressure ventilation are analyzed and reported here. Special emphasis is given to how the cyclic variation in V_c is affected by the oscillatory airway pressure and its frequency. Since the airway pressure is a part of the transmural and transpulmonary pressures that act to deform the pulmonary capillaries, this assessment provides a measure of the deformation of the pulmonary capillaries by the airway pressure for the quantification of their viscoelastic properties (Evans et al., 1987; Lee and Lee, 1986). The incorporation of this assessment into those on the mechanics of the entire pulmonary vasculature could provide a more realistic vascular model for analyzing the pressure-flow relation of the lung and the effect of ventilation on cardiac function (Brower et al., 1985; Dawson et al., 1977; Fung, 1985; Linehan et al., 1986; Yen et al., 1983).

Density Measurements and Ventilation Experiments

The density measuring system (DMS) is composed of a DMA 602 density meter (Paar Inc., Graz, Austria), a Lab Master board, and a personal computer. The blood is made to flow through the hollow U tube of the meter at a constant rate. The electronics of the meter induce the U tube to oscillate and find a frequency at which the blood-filled U tube resonates. Then the computer determines the frequency by counting the time for a given number of oscillations. Since the mass of blood in the U tube is one determinant of the resonant frequency, the computer uses previous calibration in converting the frequency measurement to the density of blood in the U tube (ρ). The third to sixth significant digit of the density measurement is then converted to analog signal for display over the next sampling period. The larger the number of oscillations the higher is the resolution of the DMS and the slower the sampling rate. For the density fluctuation experiments, we choose the number of oscillations to be 40–50. The resulting sampling rate is about five to six samples per second and the resolution is 0.03 g/l (Gamas and Lee, 1985).

The dynamic testing of the DMS is done by rapidly flushing the U tube filled with fluid of one density with another fluid of a different density. By referencing the pressure disturbance induced by th flushing with the density signal, we estimate that the time for the density to reach 95% of the expected density change is less than 0.4 s (Gamas and Lee, 1985). This result shows that the DMS is adequate to measure a variation with a frequency less than 40 cycles per minute.

The red cells have a density (ρ_r) about 65 g/l higher than that of plasma (ρ_p). Since the blood consists mainly of red blood cells and plasma, the density of blood (ρ) relates to its hematocrit (H) and the densities of its two components by the following equation:

$$\rho = \rho_p + (\rho_r - \rho_p)H \tag{10.1}$$

In our experiments, a 5-mm Hg fluctuation in the airway pressure is not likely to induce a cyclic filtration (from the vascular compartment to the tissue of

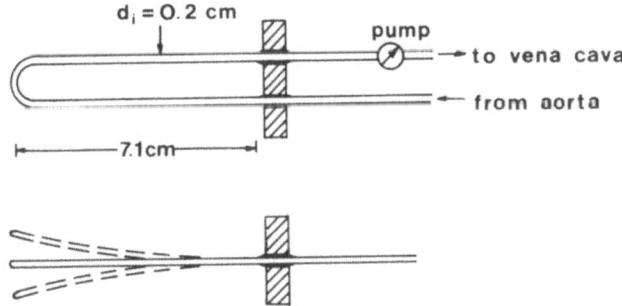

FIGURE 10.1. The setup to measure blood density. The blood from the aorta is withdrawn to flow through the U tube of the density meter and then reinfused back to the circulation. The cantilevered U tube is made to oscillate in the way depicted in the lower configuration. The frequency at which the oscillation is maximum is taken as the resonant frequency and used for the determination of the density of blood in the U tube.

the lung) at the ventilation frequency. Thus, we can consider the plasma density as a constant not affected by the ventilation.

Although a 1°C change yields a density of 0.4 g/l, it is not a problem because the animal maintains its blood at a very stable temperature. Because the volume of the glass U tube could not be changed by the fluctuation in the arterial pressure, the density measurement is not affected by the pressure variation transmitted from the aorta to the U tube. Accordingly, any density variation in the blood flowing from the lung reflects primarily a change in the hematocrit. In accordance with the resolution of the density measurement, the DMS could detect a hematocrit change as small as 0.05%.

Dogs or rabbits are anesthetized with sodium pentobarbital and infused with heparin (1,500 U/kg) intravenously. The animal is ventilated by a respirator set to provide a ventilation of constant frequency (f, cycles per minute, or ω, radian per second). As shown in Figure 10.1, a Masterflex pump withdraws the aortic blood from the carotid artery through the U tube and reinfuses it back to the animal via its jugular vein. A high flow about one-tenth of the cardiac output is selected to minimize the attenuation of the density variation due to the dispersion in the sampling catheter while not imposing excessive stress onto the cardiovascular system. Because the blood is exposed only to the tubing and the vibrating U tube, there is no blood loss or damage.

Theory of Density Dilution

The injection of a volume V_i of saline having a density (ρ_s) that is lower than the average density of circulating blood (ρ_a) into the jugular vein results in a transient decrease in the density of blood sampled from the aorta (ρ). An inverted density curve is shown in Figure 10.2A. Let us take the density difference $\rho_a - \rho_s$ as the concentration of a "density" indicator in the saline and $\rho_a - \rho$ as its concentration in the aortic blood. We could equate the "mass" of density indicator injected, $V_i(\rho_a - \rho_s)$, with the mass out of the indictor (integrated from its concentration in the aortic blood) to compute the cardiac output $\dot{Q}$. This computation has been shown to be comparable to that calculated from the dilution of Cardio-Green dye (Gamas and Lee, 1985). A correlation of the mean transit time (t_m) of the saline density indicator with that of Cardio-Green yields a relation to estimate the mean transit time of blood from the injection site to the sampling site (the U tube).

For the sequence of experiments reported here, we find that the average cardiac output as measured by saline density dilution in 12 rabbits is 320 ± 23 ml/min and the mean transit time for saline to flow from the injection site to the sampling site 5.1 ± 0.16 s. Corrected for the catheter delays and the difference between the transport of saline indicator and that of blood, the mean transit time of the blood through the central circulation, T_m, is 3.6 s.

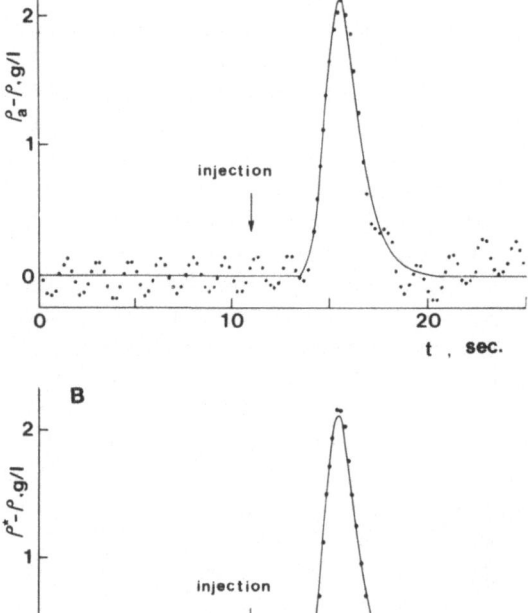

FIGURE 10.2. Saline dilution data and their fitted solid curve in the form of a lagged normal density function. Each dot is one density measurement. In graph A, we plot the difference between the average density (ρ_a) and the density of the aortic blood (ρ) to invert the dilution curve to one of conventional form. With $\rho*$ as the Fourier series representation of the periodic density variation, the display of $\rho* - \rho$ in graph B removes the cyclic variation from the saline dilution curve.

Estimating the blood volume in the left and right heart of the rabbit as 6 ml, we obtain a pulmonary blood volume of about 13 $[(360 \times (3.6/60)) - 6]$ ml. If 50% of this volume is in the capillaries (Lee, 1986), we have a V_c of 6.5 ml.

Deformation of Pulmonary Capillaries and Density Variation

The tube hematocrit of blood flowing in minute vessels is the volumetric percentage of red cell if the flow were frozen and the discharge hematocrit is that of blood discharged from the vessels. The measurement of these two hematocrits for blood flow in minute glass tubes indicates that the ratio of the tube hematocrit to the discharge hematocrit has a value less than unity (the Fahraeus effect) and that the hematocrit ratio is at its minimum when the tube diameter is comparable to that of pulmonary capillaries (Albrecht et al., 1979). If this dependence of the ratio on the diameter is applicable to the blood flow in pulmonary capillaries, then the small deformation of pulmonary capillaries as induced by ventilation will not affect the hematocrit ratio. Because no red cells are trapped by the pulmonary circulation, we could take the hematocrit of blood feeding into or discharging from the pulmonary capillaries to be the hematocrit (H_a) in large arteries or veins. Accordingly, we approximate the tube hematocrit (H_c) in the capillaries as a constant for the whole ventilation cycle.

Suppose the blood flows into the capillaries at a rate $\dot{Q}$. Over a small time increment dt, the capillary volume changes from V_c to $V_c + dV_c$. Since the volume of blood flowing into the capillaries over this time increment is $\dot{Q}\,dt$, the volume of blood out of the capillaries is, therefore, $\dot{Q}\,dt - dV_c$. Over the same time increment, the volume of red cells flowing into the capillaries is $\dot{Q}H_a\,dt$ and the increment of red cell volume in the capillaries is $H_c\,dV_c$. Thus the volume of red cells out of the capillaries is $\dot{Q}H_a\,dt - H_c\,dV_c$. Dividing the red cell volume leaving the capillaries by its corresponding blood volume, we obtain the hematocrit of the discharged blood as

$$H' = (\dot{Q}H_a - H_c\,dV_c/dt)/(\dot{Q} - dV_c/dt) \qquad (10.2)$$

Assuming dV_c/dt to be much smaller than $\dot{Q}$, we reorganize and then linearize the equation above to the following form

$$H_a - H' = [(-dV_c/dt)/\dot{Q}](H_a - H_c) \qquad (10.3)$$

As one can see, a cyclic variation in the pulmonary capillary blood volume induces a hematocrit or density variation in the blood flowing from the capillaries. The variation is proportional to the hematocrit difference between the blood in the capillaries and the blood circulating in large blood vessels.

When we reduce the pressure perfusing the lung, dV_c/dt will be transiently negative. Eq. 10.3 suggests a transient decrease in H' from H_a. To illustrate graphically this decrease, we depict an idealized distribution of hematocrit for a three-compartment pulmonary vasculature (Fig. 10.3). With the reduction in the volume of the capillary compartment, part of the low-hematocrit blood moves to the veins and induces a transient density decrease there. This density decrease has been observed in the blood flowing from an isolated lung when its arterial and venous pressure are lowered to a new level (Lee et al., 1985). The application of this theoretical consideration Eq. 10.3 to the experimental data of pressure reduction enables us to estimate a pulmonary capillary hematocrit (H_c) that is comparable to those obtained by indicator dilution techniques (Aarseth et al., 1983; Lee, 1986).

The morphometric measurements of the pulmonary vasculature indicate that the blood volume in the arterioles and venules of the lung is about 5% of V_c. Because of their larger size, the Fahraeus effect there is less pronounced than that in the capillaries. Accordingly, the density variation due to the deformation of these larger vessels will be small. Therefore, we regard the hematocrit variation sampled by the DMS as originating primarily from the deforming pulmonary capillaries (Lee and Lee, 1986).

For the Fahraeus effect in the pulmonary vasculature described in Figure 10.3, we could consider its capillary compartment as one having the hematocrit H_c and the arterial and venous compartment as one having the hematocrit H_a. Representing the total blood volume in the lung as V_b, we have the red cell volume in this three-compartment lung as $\bar{V}_c H_c + (V_b - \bar{V}_c)H_a$. According to the definition of the whole lung hematocrit (H_w), the red cell volume in the lung is also $V_b H_w$. The equality of these two volumes can be reorganized into

$$\bar{V}_c/V_b = (H_a - H_w)/(H_a - H_c) \tag{10.4}$$

Let ρ_w be the density of blood calculated for the whole lung hematocrit. The combination of Eqs. 10.1, 10.3, and 10.4 yields the following density variation

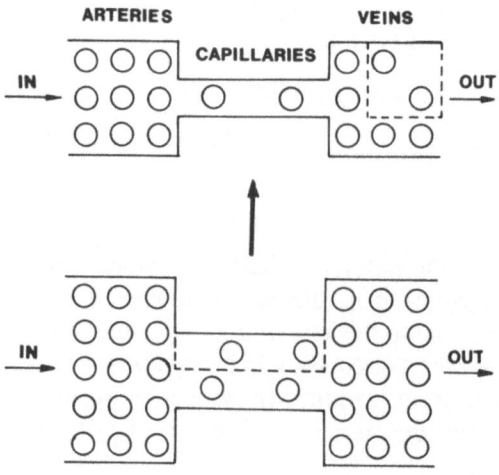

FIGURE 10.3. A diagrammatic sketch showing the distribution of red cells in a three-compartment pulmonary vasculature. A wider spacing of red cells for the capillary compartment indicates the Fahraeus effect. When the volume of the capillaries is reduced, the configuration changes from the bottom form to the top form, with part of the low-hematocrit blood moving into the venular compartment. As a result a transient density decrease is induced in the out-flowing blood. (From Lee & Lee, 1988.)

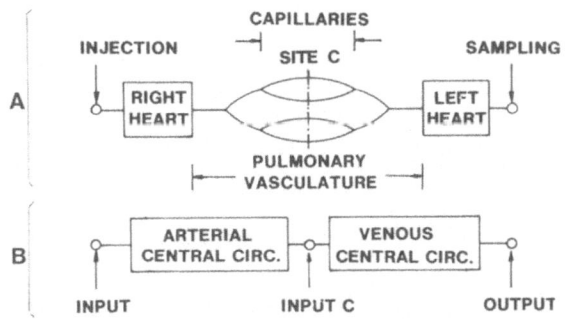

FIGURE 10.4. A two-compartment model for fractionalizing the dispersion in the central circulation into a venous dispersion and an arterial one. In the conventional study of indicator dilution, the vena cava is the injection site and the aorta the sampling site. As the capillaries are deformed by the ventilation, a cyclic variation in the density of blood is introduced into the capillary blood flow. We extend the dilution theory to this case by regarding a cyclic input being "injected" at a site C midway in the pulmonary capillaries.

for blood leaving the pulmonary capillaries

$$\rho_a - \rho' = [(-dV_c/dt)/\bar{V}_c](V_b/\dot{Q})(\rho_a - \rho_w) \tag{10.5}$$

Since the airway pressure is uniformly transmitted to all alveoli to deform the capillaries, we consider the density variation (Eq. 10.5) as one being simultaneously introduced to site C within all alveolar capillaries (Fig. 10.4). (The possible location for site C is discussed later.) As the blood transports this variation through the compartment on the venous side of C to the sampling site, the variation predicted by Eq. 10.5 is dispersed to the measured density variation.

Harmonic Representations of the Variations

In the case in which the input variation is harmonic, the computation of the convolution is simple. In addition, the sampled density variation consists mainly of a harmonic variation with ω as its frequency. Thus, we represent all variations in the following complex forms:

$$P_A - \bar{P}_A = (\Delta P/2) \exp(-i\omega t) \tag{10.6}$$

$$\rho_a - \rho = (\Delta\rho/2) \exp[-i\omega(t - t_d)] \tag{10.7}$$

$$\rho_a - \rho' = (\Delta\rho'/2) \exp[-i\omega(t - t_1)] \tag{10.8}$$

$$V_c - \bar{V}_c = (\Delta V_c/2) \exp[-i\omega(t - t_1) + i\pi/2] \tag{10.9}$$

where $\bar{P}_A$ is the mean airway pressure. The maximum changes over one ventilation cycle or the fluctuations, ΔP, $\Delta\rho$, and the time shift t_d derived from the first harmonics of the Fourier series representing the variations of the airway pressure and sampled density. The maximum changes $\Delta\rho'$ and ΔV_c, and the time shift t_1 are to be determined later by a dispersion analysis. The relation between the density and volumetric variation (Eq. 10.5) leads to the addition of $i\pi/2$ to the phase shift of the pulmonary capillary blood volume.

Two-Compartment Dispersion Model

Let h_v be the transfer function (the time concentration curve due to a pulse injection at site C) describing the dispersion of the indicator through the venous compartment. For a linear transport system, the convolution of h_v with the density variation $(\rho_a - \rho')$ yields the density variation $(\rho_a - \rho)$ being sampled by the DMS.

We let the transfer function of the compartment on the arterial side of site C be represented by h_a. By considering the arterial and venous compartments to be connected in series, the transfer function of the whole central circulation is the convolution of h_a and h_v. Since a dispersion analysis for harmonic representations has a simpler operation in the domain of Fourier transform, we designate $h_v(\omega)$, $h_a(\omega)$, and $h(\omega)$ as the Fourier transforms of the corresponding transfer functions.

Suppose the arterial network of the lung is a mirror image of its venous network. Then the dispersion of indicator in both networks is similar (i.e., $h_a = h_v$). As a result, $h_v(\omega) = h(\omega)^{1/2}$. This simple form leads us to introduce a power fraction q by taking $h_v(\omega)$ as $h(\omega)^q$ and $h_a(\omega)$ as $h(\omega)^{1-q}$. When q is bigger than 0.5, the fractionation describes a venous dispersion that is larger than the arterial one. With this fractionation procedure, the derivation of the venous transfer function is reduced to the determination of q and $h(\omega)$.

Transfer Function of the Central Circulation

With the saline density dilution curve identified as the transfer function, we fit it with a lagged normal density function, the convolution of the following exponential decay function and normal density function

$$h(t) = (1/b)\exp(-t/b)\mathbf{1}(t) * \exp\{-[(t - t_m + b)/s]^2\}/[s(2\pi)^{1/2}] \quad (10.10)$$

where $\mathbf{1}(t)$ is the unit step function, t_m is the mean transit time, and b and s are time constants (Bassingthwaite et al., 1966).

In a plot of $\rho_a - \rho$ versus t, the transient density reduction due to the injection of a saline bolus into the vena cava is inverted to a form similar to a conventional dilution curve. Figure 10.2A shows a typical density dilution curve so converted. The data presented in Figure 10.2B represent a curve with the cyclic density variation subtracted. After the recirculation was removed from this revised curve, the first, second, and third moments were computed and used to determine the three time constants in Eq. 10.10. We then calculated the lagged normal density function from the convolution and plotted it as the solid line in Figure 10.2A and 10.2B. Its fit with the measured samples has a coefficient of variation of 5%. Its mean and standard deviation for 35 dilutions is $9.0 \pm 0.5\%$.

The Fourier transform of the lagged normal density function can be expressed in the form $A \exp(i\omega t_d{}^*)$ with

$$A = \exp[-(\omega s)^2/2)]/[1 + (\omega b)^2]^{1/2} \tag{10.11}$$

$$t_d{}^* = t_m - t_r \tag{10.12}$$

where $t_r = b - (1/\omega) \tan^{-1}(\omega b)$. Eq. 10.12 indicates that the sum of t_r and the delay time between the sinusoidal input and output is the mean transit time for transporting the blood from the input to the output site. Applying this transform to the convolution of the transfer function with a harmonic input introduced to the injection site, we find that the attenuation of the input amplitude by A and a delay in the time by $t_d{}^*$ form the harmonic output.

Based on the previous fractionalization of the transfer function for the two-compartment model, we obtain the following relation to compute the attenuation of the density variation from site C to the sampling site

$$\Delta\rho = A^q \Delta\rho' \tag{10.13}$$

As one can see A^q describes the attenuation of an input oscillation with the fluctuation $\Delta\rho'$ to an output oscillation with the fluctuation $\Delta\rho$. The dependence of A on ω indicates that the larger the frequency the higher is the attenuation. This is the reason that we omit the higher harmonics in the representations of volumetric variation, density, and the like.

The time lag of the sampled density from that of the density variation generated from the capillaries is $qt_d{}^*$. In analogy to the relation between the mean transit time and the delay time (Eq. 10.12), we add qt_r to the time lag to form an estimate of the mean transit time for the blood to traverse the venous compartment:

$$t_v = qt_d{}^* + qt_r = qt_m \tag{10.14}$$

The Quotient q and the Venous Mean Transit Time

This two-compartment analysis and our previous observation on the time delay t_d led to a procedure to assess the quotient q. Demonstrated by our ventilation experiments with a wide frequency spectrum, the mean transit time t_m is independent of the ventilation frequency. We find the addition of qt_r to t_d (the delay time of the density variation from the airway pressure) yields a time that is not dependent on the frequency. The value of this time is consistent with a mean transit time expected for the blood to flow from a site midway of the pulmonary capillary to the sampling site. These results lead us to identify

t_d as qt_d^*. (This identification of t_d sets the t_1 in Eq. 10.8 as zero). By correlating t_d with t_d^* for 35 dilution experiments (Fig. 10.5A), we obtain a q of 0.52 with a correlation coefficient of 0.94.

Using this q for all dilution experiments, we determine the venous mean transit time t_v and plot it against t_m in Figure 10.5B. A linear relation with q as the slope is indicated. Included in the figure are the times measured from the dog's experiments. Subtracting the catheter delay in a way similar to the computation of the mean transit time T_m, we find that the venous mean transit time T_v (for blood to traverse from site C to the pulmonary veins) normalized by T_m is 0.44 ± 0.01.

Grouping the data to a ventilation frequency of 20 ± 1, 25 ± 1, 30 ± 1, or 35 ± 1 cycles per minute, the correlation of t_v and t_m of each data group yields a q of 0.48, 0.51, 0.51, or 0.53, respectively. The small deviation in q reconfirms the previous assumption that t_v should not be dependent on the ventilation frequency.

Many investigators have monitored the appearance of tracer gases in the

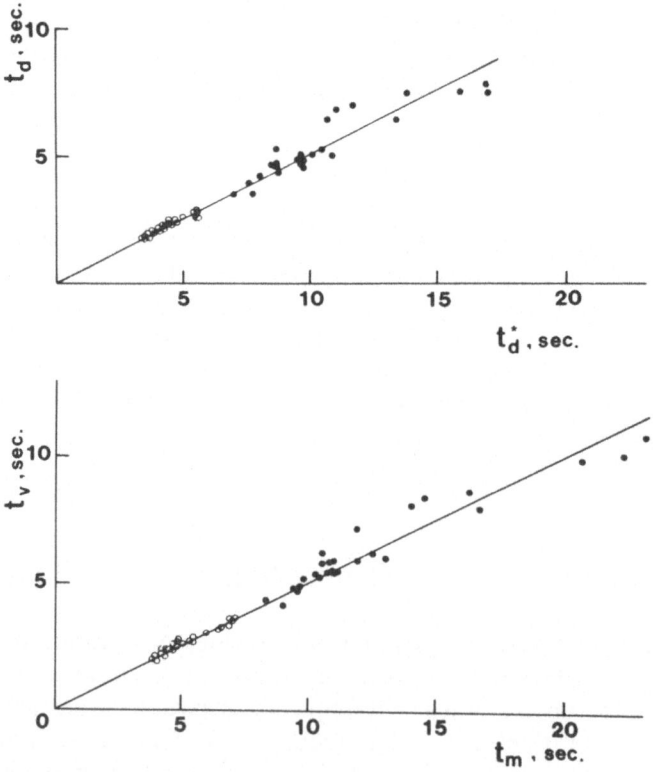

FIGURE 10.5. A: Correlation between the time delay (t_d) of the density variation from the airway pressure and the mean transit time t_m of the rabbit's central circulations. B: Correlation between the mean transit time (t_v) for the venous compartment and t_m.

alveoli for the determination of the arterial-to-capillary-to-venous fractional volumes of the pulmonary vasculature. If the venous ends of the capillaries were site C in the two-compartment analysis, a time ratio T_v/T_m of $30 \pm 3\%$ (for five independent studies reviewed in Lee, 1986) would be obtained. On the other hand, a site at the arterial ends would lead to a time ratio of $73 \pm 2\%$. Our result for T_v/T_m (44%) suggests that site C is near midway in the pulmonary capillaries.

The density indicator introduced near the arterial end of capillaries requires a transit time longer than that introduced near the venous end. In an analysis of the motion of red cells in deforming capillaries, we find that the density variation can be considered as one introduced midway in the pulmonary capillaries (Friend and Lee, 1989).

Recently, Dawson et al. (1987) reported that the transit time for blood to travel from the main pulmonary artery to the subpleural arterioles (20–30 μm) on the dependent surface of the dog's lungs is 1.94 s. Judging from their diameter, this transit time may be regarded as that from the main pulmonary artery to the entrances of the capillaries. The mean transit time through the subpleural capillaries on the dependent surface of the lung has been measured as 1.75 s (Capen et al., 1987). Accordingly, the mean transit time from the main pulmonary artery to the midpoints of the pulmonary capillaries is 2.82 s. Estimating the mean transit time of the vasculature (T_m) of these dog's lungs as 5 s (Gamas and Lee, 1985), we obtain a venous transit time (from the midpoints of the capillaries to the pulmonary veins) of 42% of T_m. This is comparable to a T_v/T_m of 45% found by the density method for dog's central circulation (Lee, 1988).

Volumetric Deformation of Pulmonary Capillaries

The density fluctuation in aortic blood for 12 rabbits ($\Delta\rho$) is plotted against the frequency in Figure 10.6A. It shows a decreasing density fluctuation as the frequency increases. From Eq. 10.13, we calculate the density fluctuation of the blood leaving the capillaries. As shown in Figure 10.6B, $\Delta\rho'$ is approximately a constant not affected by the frequency.

The combination of Eqs. 10.5, 10.8, 10.9, and 10.13 yields the following equation to calculate the volumetric fluctuation

$$\Delta V_c/\bar{V}_c = [\Delta\rho/(\rho_a - \rho_w)](\omega T_m A^q) \tag{10.15}$$

where $T_m = V_b/\dot{Q}$. In this equation, $\Delta\rho$ is determined from the measured density variation, the attenuation A from the fitting of the dilution curve, and the mean transit time T_m from the measurement of t_m. The value of q is chosen as 0.52 and the density difference $\rho_a - \rho_w$ is computed from the assumption that H_w is 90% of H_a (Aarseth et al., 1983; Lee, 1986). The maximum variation in the rate change $|dV_c/dt|$ ($= \omega\Delta V_c/2$) and ΔV_c are shown in Figure 10.6C and

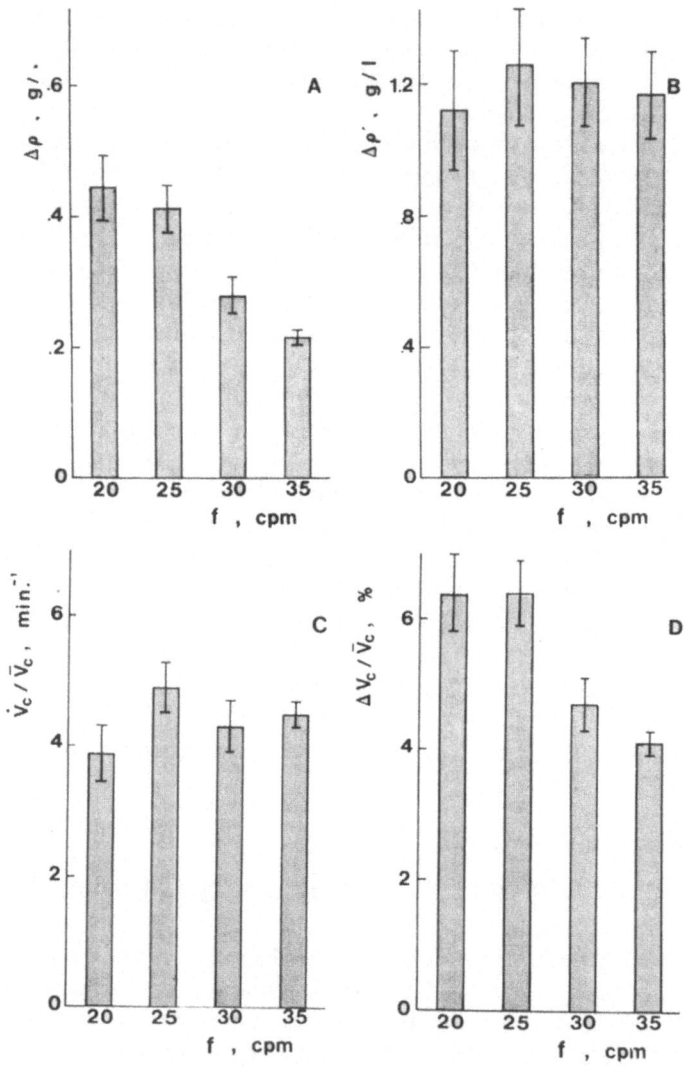

FIGURE 10.6. *A*: Density fluctuations of aortic blood ($\Delta\rho$) within one ventilation cycle as a function of the ventilation frequency f (cycles per minute; cpm). The rabbits were ventilated by a 5-mm Hg end-inspiratory pressure. *B*: Density fluctuations estimated for the blood leaving the pulmonary capillaries ($\Delta\rho'$). *C*: Fluctuations of $|dV_c/dt|$ computed by the deformation analysis. *D*: Fluctuations in the blood volume of pulmonary capillaries ($\Delta V_c/\bar{V}_c$) within one ventilation cycle.

10.6D. In these rabbit experiments, the fluctuation in the airway pressure is 5 mm Hg for all frequencies.

The normal ventilation frequency for the rabbit (about 35 cycles per minute) is higher than that of the dog (about 15 cycles per minute); we find that the density change within one ventilation cycle at the normal frequency of these two species is comparable and has a value of 0.22 g/l (Lee and Lee, 1986). It is interesting to note that their percentage change in V_c within a ventilation cycle is about the same (4%). Even though the mean transit time for rabbits is one-third to one-half of those for dogs, the comparability in $\Delta V_c / \overline{V}_c$ results from similar attenuation because the values of ωt_m, s/t_m, b/t_m, and q are comparable for the two species.

Discussion

If the filtration were an important factor leading to the density fluctuation for the rabbits, then our previous estimate of the rate change of V_c should now be interpreted as a cyclic variation in the rate of filtration. Suppose that the ventilation induces a net cyclic driving pressure with an amplitude of 2.5 mm Hg (that of the airway pressure). With the net tissue weight of a lung being approximately its blood volume, we estimate that the filtration requires the lung to have a capillary filtration coefficient as high as 47 ml/h/mm Hg/g of wet tissue. Since the filtration coefficient for isolated rabbit lungs is only 0.1–0.6 ml/h/mm Hg/g of tissue (Wangensteen et al., 1977), we conclude that the ventilation could not generate a cyclic filtration large enough to account for the aortic density fluctuation in the rabbit.

A cyclic recruitment of the pulmonary capillaries induced by the ventilation is not a likely explanation of the density fluctuation. Because the in vivo experiment is done under the supine condition, the capillary blood flow is probably under a zone 3 condition. Thus the distensibility of the pulmonary capillaries play a more important role than their recruitment in changing the blood volume of capillaries with flow. We have also observed the same density fluctuation when we switch the rabbit in a box for an intermittent negative box pressure ventilation (Lee et al., 1988). In contrast to an intermittent positive-pressure ventilation, the negative one will definitely assure a zone 3 capillary blood flow.

To time the events occurring in the pulmonary capillaries, we display the density variation in the aortic blood ($\rho_a - \rho$), the airway pressure, the density variation of blood leaving the capillaries ($\rho_a - \rho'$), $[dV_c/dt]/\overline{V}_c$, and $V_c/\overline{V}_c$ as a function of the time in Figure 10.7. The last three are computed from the dispersion analysis. As one can see, the minimum density occurs at the same instant as the maximum airway pressure and the variation in V_c is inphase with the airway pressure. Because the blood flow out of the lung will be affected by this volumetric change, its role in the filling of the left heart is one facet of how ventilation affects the cardiac function.

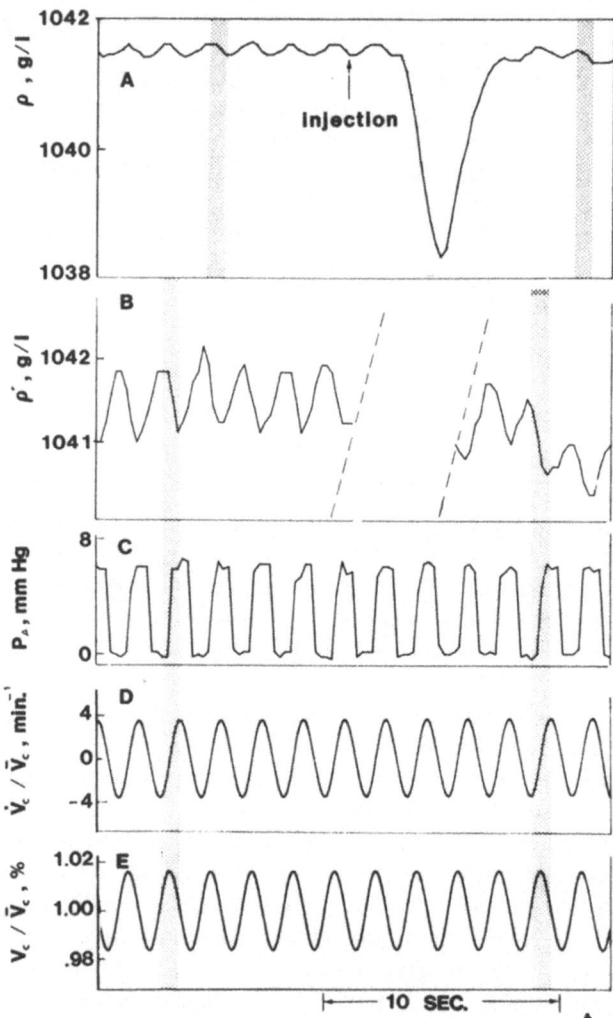

FIGURE 10.7. The variations in: A, the density (ρ) of aortic blood; B, the density (ρ') of blood leaving the pulmonary capillaries (this variation is that in panel A divided by an attenuation A^q and shifted by a time delay t_d); C, the airway pressure P_A; D, the rate change of the pulmonary capillary blood volume ($[dV_c/dt]/\bar{V}_c$); and E, the normalized blood volume ($V_c/\bar{V}_c$). The last two are the responses derived from the first harmonic variation of the density (ρ'), and the transient decrease shown in A results from the injection of a saline bolus into the vena cava at the time marked by the arrow. The shaded strips serve as a reference for correlating these time-varying events during the inspiration period.

Let us visualize the effect of the airway pressure on the pulmonary capillaries by modeling them as an alveolar sheet (Fung, 1984). For intermittent positive-pressure ventilation, the airway pressure is transmitted to the alveoli to become a part of the transmural pressure that distends the sheet. An increase in the airway pressure decreases the sheet thickness and hence the pulmonary capillary blood volume. On the other hand, the increase in the airway pressure leads to an increase in the transpulmonary pressure, which expands the lung to increase the surface area of the sheet and hence the capillary blood volume. The combination of these two contrasting effects of the airway pressure on V_c produces certain capillary deformation leading to the density variation in the aortic blood. A mechanical simulation of these two factors suggests that the deformation is derived primarily from the viscous part of the viscoelastic property of the pulmonary capillaries (Evans et al., 1987).

In summary, the quantification of the change in V_c by the microinvasive density method and attenuation analysis is a first step to establish the relation between ventilatory loading and the deformation of the pulmonary capillaries. This relation is useful to characterize the viscoelastic properties of the pulmonary capillaries.

Acknowledgment. This research is partially supported by grants HL-36285 and HL-40893 from the National Institute of Health.

References

Aarseth P, Waaler BA, Nicolaysen G (1978) Rabbit lung plasma and erythrocyte volumes, lung hematocrit in relation to total body hematocrit. *Acta Physiol Scand* 103:165–172.

Albrecht KH, Gaehtgens P, Pries A, Heuser M (1979) The Fahraeus effect in narrow capillaries (i.d. 3.3 to 11.0 μm). *Microvasc Res* 18:33–47.

Bassingthwaighte JB, Ackerman FH, Wood EJ (1966) Applications of the lagged normal density curve as a model for arterial dilution curves. *Circ Res* 18:398–415.

Brower R, Wise RA, Hassapoyannes C, Bromberger-Barnea B, Permutt S (1985) Effect of lung inflation on lung blood volume and pulmonary venous flow. *J Appl Physiol* 58:954–963.

Capen RL, Latham LP, Wagner WW Jr (1987) Comparison of direct and indirect measurements of pulmonary capillary transit times. *J Appl Physiol* 62:1150–1154.

Dawson CA, Capen RL, Latham LP, Hanson WL, Hofmeister SE, Bronikowski TA, Rickaby DA, and Wagner WW Jr (1987) Pulmonary arterial transit times. *J Appl Physiol* 63:770–777.

Dawson CA, Grimm DJ, Linehan JH (1977) Effects of lung inflation on longitudinal distribution of pulmonary vascular resistance. *J Appl Physiol* 43:1089–1092.

Evans MV, Lee JS, Lee LP (1987) Time shift in ventilation induced density fluctuation in arterial blood. *Ann Biomed Eng* 19:234–255.

Friend M, Lee JS (1989) Red blood cell motion in capillaries subject to cyclic deformation. *FASEB J* 48:A1386.

Fung YC (1984) *Biodynamics: Circulation*, Springer-Verlag, New York.

Gamas L, Lee JS (1985) Density indicator method to measure pulmonary blood flows. *J Appl Physiol* 60:327–334.

Lee JS (1986) Microvascular hematocrit of the lung. In Schmid-Schonbeim G, Wu S, and Zweifach BW (eds) *Frontiers in Biomechanics*, Springer-Verlag, New York pp. 353–364.

Lee JS (1988) Assessing the deformation of pulmonary capillaries produced by ventilation. *Microvasc Res* 35:48–62.

Lee JS, Fallon T, Hunter M, Ye Q, Lee LP (1988) Respiratory effect on the blood volume of pulmonary capillaries. *J Biomech Eng* 110:150–154.

Lee JS, Lee LP (1986) Ventilatory changes of pulmonary capillary blood volume assessed by arterial density. *J Appl Physiol* 61:1724–1731.

Lee JS, Lee LP (1988) Effect of hemodilution on ventilatory fluctuations of pulmonary capillary blood volume. *J Appl Physiol* 65:2571–2578.

Lee JS, Lee LP, Evans MV, Gamas L (1985) A density method to quantify pulmonary microvascular hematocrit. *Microvasc Res* 38:222–234.

Linehan JH, Dawson CA, Rickaby DA, Bronikowski A (1986) Pulmonary vascular compliance and viscoelasticity, *J Appl Physiol* 61:1802–1814.

Piiper J (1971) Attempts to determine volume, compliance and resistance to flow of pulmonary diffusion capacity, IV. The normal dog lung. *Resp Physiol* 13:141–159.

Roughton FJW, Forster FE (1957) Relative importance of diffusion and chemical reaction rates in determining rate of exchange of gases in the human lung, with special reference to true diffusing capacity of pulmonary membrane and volume of blood in the lung capillaries. *J Appl Physiol* 11:290–302.

Wangensteen OG, Lysaker E, Savaryn P (1977) Pulmonary capillary filtration and reflection coefficients in the adult rabbit. *Microvasc Res* 14:81–97.

Yen RT, Zhung FY, Fung YC, Ho HH, Tremer H, Sobin SS (1983) Morphometry of cat's pulmonary venous tree. *J Appl Physiol* 55:236–242.

11
Direct Measurement of Pulmonary Capillary Transit Time and Recruitment

Wiltz W. Wagner, Jr.

Introduction

Capillary recruitment and transit time in the lung are important determinants of blood oxygenation, especially during exercise. Significant gaps exist, however, in our understanding of the way in which alterations of pulmonary hemodynamics affect capillary transit time and recruitment in various regions of the lung. Part of the reason for this lack of information is the considerable technical difficulty in studying pulmonary microvessels directly. The classical and most direct approach is in vivo microscopy, but tissue movement during the cardiorespiratory cycles is a substantial impediment to data collection. Despite the technical difficulties, historic discoveries have come from using this method.

Early Microvascular Observations

In 1661 Malpighi made the first in vivo microscopic observations of the surface of the lung using an early compound microscope (Malpighi, 1661). His efforts resulted in two great discoveries, the first of which was red blood cells. With the second discovery, that of capillaries, he solved the millennia-long problem of how the arteries and veins were connected. It is tragic that Malpighi's work came four years after Harvey's death, for Harvey never solved the vexing anatomical problem of how the blood actually crosses from artery to vein.

In the next century, Hales made the first measurements of capillary transit time (Hales, 1733). He reported that the red cells moved through the capillaries at the rate of "1/10th of an Inch in the Time of eight beats of a Watch which beat 16,000 times in an Hour." This converts to 1,200 μm/s, a figure in close agreement with modern measurements.

In this century, Wearn et al. (1934) made observations of the cat lung through a window in the parietal pleura. From the observation that not all of the capillaries were perfused during rest, important conclusions were drawn about the mechanism of gas exchange reserve in the lung. In the early 1960s, Giles Filley instigated work on the pulmonary microcirculation using in vivo microscopy (de Alva et al., 1962; Krahl, 1963) in an attempt to determine which pulmonary vessels constricted during airway hypoxia. That issue was one of the hotly debated topics of the day (Grover, 1963). The argument focused on two anatomical sites. Those favoring a precapillary site reasoned that it was only the arteries that had sufficient smooth muscle in their walls to constrict, that it was the arterial muscle that hypertrophied at high altitude, and finally, that the precapillary site was the most suitable location for regulating ventilation-perfusion balance, since arterial constriction would direct blood flow to better ventilated parts of the lung before it reached hypoxic alveoli. The veno-constriction camp contended that the veins were the only vessels that could sample the oxygen levels in blood leaving underventilated alveolar regions and, therefore, the postcapillary vessels had to be the site of hypoxic vaso-constriction. In the intervening years, a variety of evidence has made it apparent that it is arteries of 200–300 μm in diameter that constrict during hypoxia (Dawson et al., 1979; Kato and Staub, 1966; Nagasaka et al., 1984; Sackner et al., 1966; Shirai et al., 1986) and that the arteries are themselves capable of directly sensing airway oxygen tension (Conhaim and Staub, 1980; Jameson, 1964; Staub, 1961).

In Vivo Pulmonary Microscopy

When Dr. Filley and I began working on a method for in vivo pulmonary microscopy, we had the good fortune of discovering a location on the surface of the lung lying under the second rib that moved only slightly with respiration (Wagner and Filley, 1965). We implanted a transparent thoracic window at that location and were able to make observations at medium magnifications in preparations with normal blood pressures, cardiac outputs, and blood gases. Later a suction manifold was added to the inside of the window frame so that movement was reduced to 10 μm or less with each breath (Wagner, 1969). This permitted us to study one field over many hours and thereby to use the same arterioles, capillaries, and venules as their own controls. The stage seemed set to discover the site of hypoxic vasoconstriction. Had we done sufficient preliminary theoretical work, we would have realized that the arterioles and venules on the surface of the lung rarely exceed 50 μm in diameter and that in our experimental animal (dogs), smooth muscle is un-common in vessels smaller than 100 μm. This combination made it unlikely that we would be able to detect vasomotion. For better or worse, we did not reason this out ahead of time. The result has been that despite 25 years of attacking the canine pulmonary circulation with every sort of vasoconstrictive abuse we could imagine, we have yet to see a microvessel constrict and so have made no direct contribution to our original goal of understanding which vessels constrict with hypoxia.

On the other hand, we have developed the ability to observe directly the

FIGURE 11.1. This curve represents the average effect of airway hypoxia on the level of capillary recruitment in the upper lung in a series of nine consecutive dogs. The curve swings upward at the same oxygen tension that systemic arterial hypoxemia begins to occur. (From Wagner and Latham, 1975.)

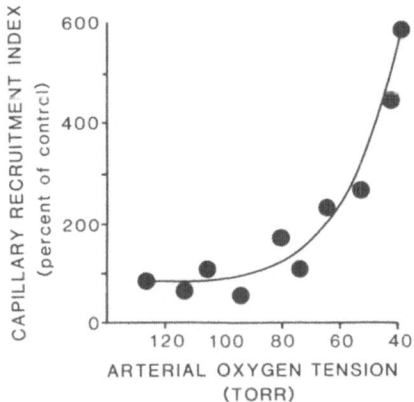

pulmonary gas exchange vessels in a still field in an animal with normal cardiopulmonary physiology and, as best we can determine, an undamaged microvasculature (Wagner, 1969; Wagner and Filley, 1965). This has enabled us to undertake a number of studies of how the pulmonary capillaries respond to changes in pressure and flow, observations that form the basis of this chapter.

The majority of our studies have been made on the surface of the left lower lobe with the animal lying in the right lateral decubitus position. The microscope is directed downward onto the uppermost surface of the lung. This position places the field in zone 2 hydrostatic conditions, where pulmonary arterial pressure > alveolar pressure > pulmonary venous pressure (West, 1979). Zone 2 is interesting because that region is where much of the gas exchange reserve lies in terms of recruitable capillaries (Wagner and Latham, 1975). In our early studies we noticed that red cells consistently perfused more capillaries when the animals inspired hypoxic gas mixtures (Wagner and Latham, 1975). To quantify this recruitment, we made drawings of all capillaries that were perfused by red blood cells in a given field during normoxia and hypoxia. By measuring the total length of those capillaries, we demonstrated that capillary recruitment increased as oxygen tension fell (Fig. 11.1).

Capillary Recruitment and Hypoxia

This observation seemed completely paradoxical and led to a series of studies designed to determine what caused the capillary recruitment. One obvious extrapulmonary cause, increased left atrial pressure, was eliminated because that pressure was unchanged by hypoxia (Wagner and Latham, 1975; Wagner et al., 1979). In later work, we held output constant from control to hypoxia and found that recruitment occurred independent of total pulmonary blood flow (Capen and Wagner, 1982; Capen et al., 1981). From these studies, an intrapulmonary mechanism seemed likely to be the cause of the recruitment.

Venoconstriction could certainly cause a retrograde rise in capillary pressure that would lead to recruitment. The well-known elevation of pulmonary arterial pressure during hypoxia might also account for the recruitment by redistributing blood flow upward toward our zone 2 observation site. To differentiate between these potential causes, we directly monitored pressure in both small pulmonary veins and arteries, made the animal hypoxic, and measured capillary recruitment. Then, while continuing the hypoxic challenge, the vasodilator prostaglandin E$_1$ was infused to relieve whatever vasoconstriction had occurred (Capen and Wagner, 1982). The vasodilator caused a large reduction in the number of perfused capillaries. We could not measure any increase in pulmonary venous pressure during hypoxia, nor detect any effect of the vasodilator on vein pressure. Pulmonary artery pressure, however, fell to near control levels during vasodilator infusion. The plots of pressure versus recruitment (Fig. 11.2) showed no correlation with venous pressure but an impressive correlation with arterial pressure. Therefore capillary recruitment during hypoxia did not correlate with cardiac output, pulmonary venous pressure, or left atrial pressure, but did correlate with pulmonary arterial pressure.

If recruitment is caused by a rise in capillary pressure, which it certainly must be, then how could the pressure rise in the capillaries that are located downstream from an upstream arterial constriction? It seemed more plausible that constriction of an artery feeding a capillary bed would have to be associated with reduced flow and derecruitment.

To help visualize the pressure, flow, and resistance relationships in the

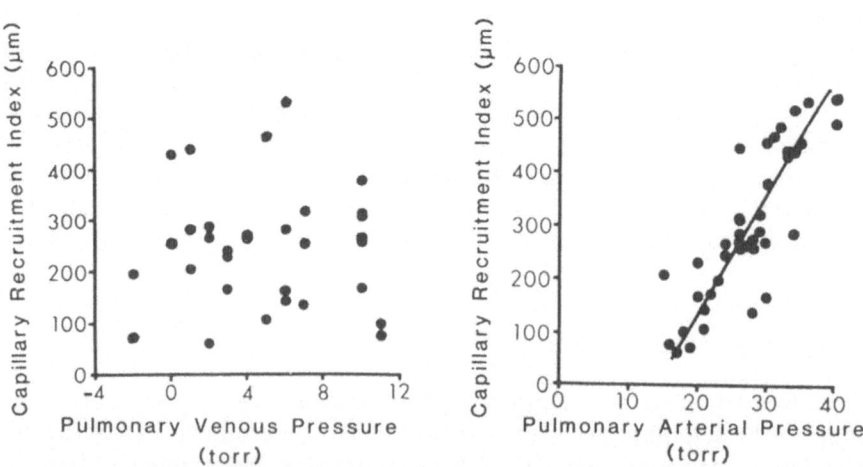

FIGURE 11.2. The correlation between capillary recruitment index and pulmonary venous pressure was not significant ($p = .9$), but was highly significantly correlated with pulmonary arterial pressure ($p < .001$) for this group of 10 dogs. (From Wagner et al., 1979.)

pulmonary circulation we developed a model. In some ways the model is undeniably too simplistic, but it has explained much of our data and has suggested a number of further experiments. The model is based on an apartment building and what happens when all of the tenants try to take a shower at 7:00 AM (Fig. 11.3, left). When everyone opens their shower control valves simultaneously, the dwellers on the lower floors benefit from a rapid flow of water through a fully recruited shower head [analogous to the capillary bed in zone 3 where pulmonary arterial pressure > pulmonary venous pressure > alveolar pressure (West, 1979)]. The upper-floor tenants are confounded by a derecruited shower head. To this point, the model reproduces in part the elegant pulmonary circulatory hydrostatic zone model developed by Permutt et al. (1965) and West (1979) to describe the hydrodynamic distribution of blood flow under normal conditions. To extend the apartment house model, let us suppose that there is a meeting in which the tenants all agree to open their control valves only partially (Fig. 11.3, right). This condition represents generalized pulmonary arterial constriction. With higher pressure available, water is redistributed to the upper floors where holes in the shower heads are recruited. The paradox is solved of how capillaries can be recruited downstream from an upstream constriction, because some water is available to flow past the constriction in the control valves to the upper floors (Fig. 11.3, right), whereas no water had been available to flow past the wide-open control valve under control conditions (Fig. 11.3, left).

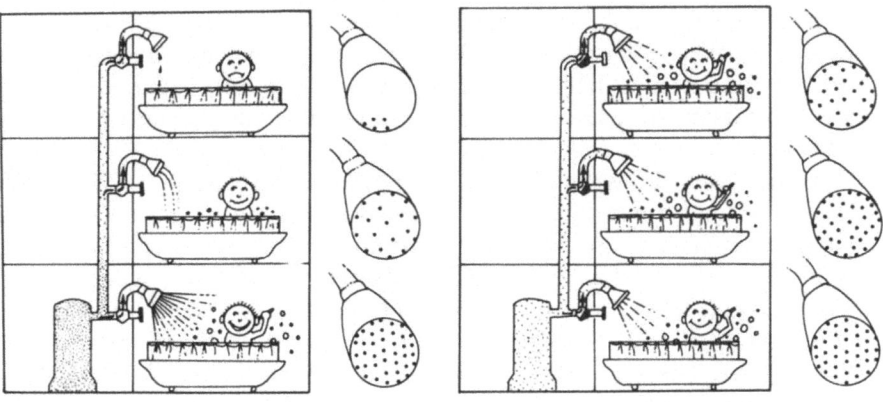

Low pulmonary arterial pressure   High pulmonary arterial pressure

FIGURE 11.3. In this model the apartment house represents the lung and the shower heads the capillary bed. Capillary recruitment is analogous to recruitment of the holes in the shower head. Under normal conditions, the shower heads are not fully recruited in the upper stories and flow is brisk in the lower stories (*left panel*). During hypoxia, arterial vasoconstriction (represented by the partially closed control valves in the *right panel*) causes pressure to be elevated and flow to be redistributed to the upper shower heads, where recruitment occurs.

The model predicts, with conditions of steady total flow (analogous to lack of the cardiac output changes during hypoxia) and even constriction in all control valves, that there will be upward redistribution of flow, which could result in an overall gain in the number of holes in the shower heads via recruitment. Under these conditions, the extra flow to the upper shower heads must come from the lower apartments. The lesser flow in the lower apartments, however, need not lead to derecruitment; rather, the extra water could come from a reduction in the velocity of the water passing through the lower shower heads (analogous to slower transit times). If this reasoning is correct, then the model predicts that in the lung there should be an increase in capillary volume during the increased pulmonary arterial pressure associated with airway hypoxia.

To test this prediction, it was necessary to determine the effect of hypoxia on total pulmonary capillary volume. To do this, we measured the diffusing capacity of the lung for carbon monoxide (Capen et al., 1981). A diffusing capacity increase during hypoxia, however, could reflect either increased capillary volume (recruitment), or less competition from oxygen for hemoglobin binding sites [which would alter the reaction rate between carbon monoxide and hemoglobin (θ)], or some combination of the two. To determine the effect of recruitment alone, the vasodilator prostaglandin E_1 was infused while airway hypoxia was held at a constant level. The resultant decrease in pulmonary artery pressure caused, as expected from earlier work (Wagner et al., 1979), capillary derecruitment (Fig. 11.4, left). Diffusing capacity also decreased (Fig. 11.4, right). By assuming no change in membrane diffusing capacity, and by having held θ constant by keeping the level of hypoxia constant, the experiment showed that hypoxia caused a net gain in capillary volume, most likely through capillary recruitment. Such an increase in gas exchange surface area would be advantageous during whole lung hypoxia.

The apartment house model predicts upward redistribution of flow with constriction and downward redistribution with dilation. There is evidence that hypoxia causes upward redistribution of pulmonary blood flow both acutely in anesthetized dogs (Dugard and Naimark, 1967) and in man nature to high altitude (Dawson and Grover, 1974). To determine what vasodilation did to blood flow distribution in our preparation, we injected radiolabeled 15-μm microspheres into the right atrium and measured the location of the wedged spheres in the pulmonary arteries during hypoxia and hypoxia plus prostaglandin E_1. As expected, hypoxia caused microsphere distribution to be relatively even from top to bottom (Fig. 11.5). The vasodilator caused the curve to rotate in a way that increased the slope, indicating that blood flow fell in the upper lung and returned to high levels in the lower lung. This seems convincing evidence that upward redistribution of blood flow existed in our preparations and is a likely explanation for capillary recruitment in the upper lung during hypoxia.

From these data, the response of the pulmonary circulation can be summarized in the following way: hypoxia $\rightarrow$ pulmonary arterial constriction $\rightarrow$

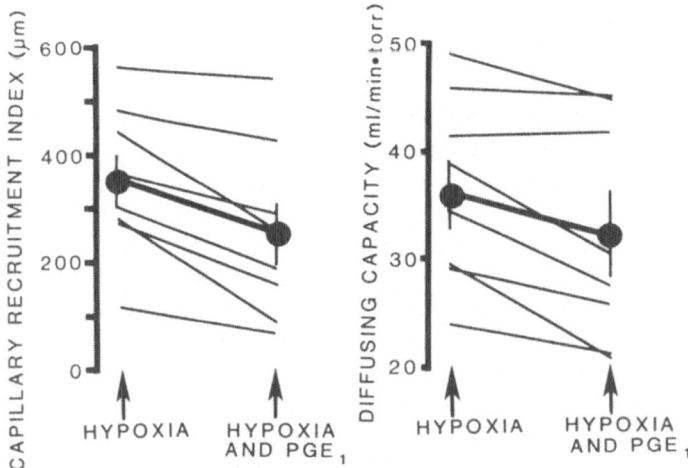

FIGURE 11.4. Capillary recruitment index and the diffusing capacity of the lung for carbon monoxide are plotted for a group of eight dogs. The thin lines are data from individual dogs. Heavy lines are group means and standard errors. Under these conditions hypoxia was held constant and recruitment was varied by infusing the vasodilator prostaglandin E_1. The derecruitment resulting from the vasodilator was associated with a fall in diffusing capacity. (From Capen, 1979.)

FIGURE 11.5. The intrapulmonary distribution of blood flow determined by radioactive microsphere distribution per gram of dry lung tissue plotted against distance up from the bottom of the lung ($n = 7$). During hypoxia the blood flow is fairly even from bottom to top, but when the vasodilator prostaglandin E_1 is infused to relieve the pulmonary arterial constriction the curve rotates, causing the slope to be steeper, an indication that blood flow is redistributed toward the bottom of the lung. The slopes of the lines are different from each other ($p < .05$). (From Capen and Wagner, 1982.)

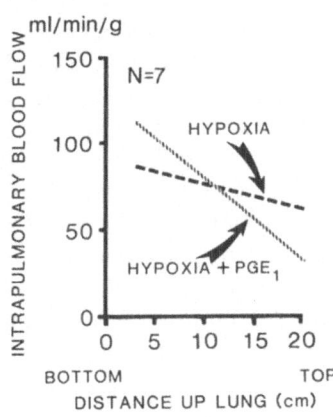

increased pulmonary arterial pressure → upward redistribution of blood flow → capillary recruitment → increased surface area for gas exchange. It is not certain how much improvement in arterial oxygen tension might occur from the increase in gas exchange surface area provided by the recruited capillaries. As best we can calculate it might not exceed 5 torr in a normal lung, although it could be substantially more in a heterogeneously diseased lung. In any case, whenever arterial oxygen tension is below 30 or 40 torr, any addition would be welcome.

Capillary Transit Time

The changes discussed thus far involve capillary recruitment, which is only part of how gas exchange surface area can be increased to oxygenate more blood in the pulmonary microcirculation. Another important variable is the length of time that the blood spends in the capillaries. There is now substantial evidence (Wagner, 1977) that red cells become oxygenated in about the first quarter of their way across the capillary bed. That means that for the majority of their passage across the gas exchange vessels, the red cells are fully oxygenated. With exercise, transit time is reduced from resting values of 1 s or so to about 0.25 s, the minimum time required for complete oxygenation of a red cell. This alteration alone permits, on average, three to four times as much blood to cross the pulmonary gas exchange vessels, become fully saturated, and thereby increase the delivery of oxygen to the body.

This picture of capillary transit time has been obtained from calculations that are based on indirect techniques and on a number of other factors, either assumed or based on other indirect measurements. It is important, therefore, to determine transit time measurements using direct methods. Further, the classical measurements provide estimates only of average red cell transit times. There is no information available about the range of transit times, either in different regions of the lung or in different parts of the capillary bed. It is possible, even likely, that there is a wide range of transit times in different parts of the lung and also in capillary networks. These variables could have a significant impact on our understanding of gas exchange, and also have potentially important clinical implications as well.

Transit times through the capillaries are even more difficult to measure than capillary recruitment. When our work started, the most viable technical approach was high-speed cinemicrography. Years of effort using this approach (Wagner et al., 1967, 1969) provided virtually no useful biological information. Even if we had succeeded, we would have only known red blood cell velocity for a few cells in a small population of individual capillaries. A more palatable approach would be one that gives information about the mean and distribution of transit times across entire capillary networks in different parts of the lung.

Those measurements became possible with the advent of very sensitive television cameras that could detect the passage of flurorescent dye through the microcirculation (Wagner et al., 1982). We used fluorescein isothiocyanate (FITC)-labeled dextran, injected through a side-hole catheter into the right atrium and recorded on videotape as the dye appeared and disappeared from the observed field. When the videotape was replayed, the video signal from a portion of the television image was electronically sampled, passed through a sample-and-hold circuit at the television field rate of 60 Hz, and recorded oscillographically. The transit of the dye across the sampled portion of the video image produced a dye dilution curve (Fig. 11.6).

To measure mean capillary transit time, it was necessary to determine from

FIGURE 11.6. Dye dilution curves obtained from a 15-μm arteriole and a 15-μm venule serving a common capillary bed. Dye is dispersed in the capillaries, causing the width of the venular curve to be greater than that of the arteriolar curve. By subtracting mean times of curves, mean capillary transit time is obtained.

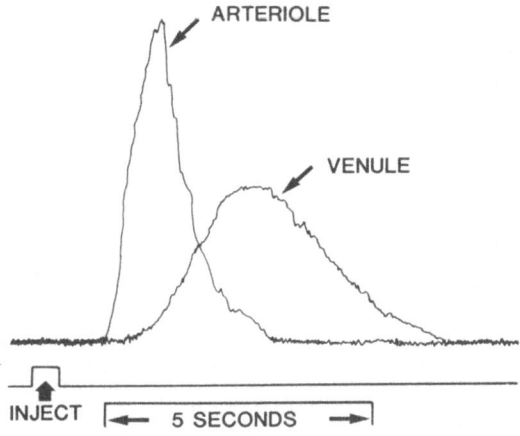

the television signal when an average particle in the dye bolus entered and exited a single capillary bed. Dye-dilution curves defining the dye passage were recorded first from a site in an immediately precapillary arterioles (less than 15 μm) and then from a site over an immediately postcapillary venule (less than 15 μm) serving a common capillary network. The time at which the average dye particle passed the sample window was determined by calculating the mean transit time of each curve. Mean capillary transit time (time-weighted mean) was the difference between transit times of the arteriolar and venular curves with references to the common injection time signal (Fig. 11.6).

Vertical Gradient of Capillary Transit Time

This technique had the potential for determining how quickly blood flows through capillaries in various hydrostatic zones, which turns out not to be a straightforward prediction. There are not many capillaries perfused by erythrocytes in the upper lung (West, 1979); this reduces the perfused cross-section of the capillary bed and may compel red cells to traverse the few patent capillaries at the 1 s time typical of the whole lung. On the other hand, the low perfusion pressure in the upper lung could cause the red cells to meander with inefficient slowness through the network. If capillary transit times are lengthy in the upper lung zones, they must be more rapid near the bases. This would create a gradient of transit times down the lung and result in simultaneously existing, gravitationally induced recruitment and transit time reserves in the capillary networks at each hydrostatic level that could be utilized to oxygenate more blood.

To determine whether a vertical gradient of capillary transit times exists, we implanted our window so that the same area of lung was studied, but the animal were oriented in different positions with respect to gravity. To study the uppermost lung surface, the animals were placed in the right lateral decubitus position. The microscope was directed downward through the window toward the surface of the "upper" lung located in zone 2. A "midlung" position was studied by placing the animals on a 20° sloped platform in the head-up, semirecumbent position. Observations were made of the side of the lung at the transition betwen zones 2 and 3. The "lower" lung was studied with the animals placed in the left lateral decubitus position so that the window was on the bottom and observations were made upward of the lowest surface of the lung, which was located in zone 3. Capillary transit times and the levels of capillary recruitment were measured in each location.

A vertical gradient did exist for capillary transit times (Wagner et al., 1986). The longest times were in the upper lung and decreased progressively toward the lower lung (Fig. 11.7, left). There was the expected distribution of capillary recruitment, with few capillaries perfused in the upper lung and highly recruited networks in the lower lung (Fig. 11.7, right). The few patent capillaries in the upper lung perfused by slowly moving red cells suggests that a potentially important reserve for gas exchange during execise lies in the lung above the heart. The reserve consists of capillary recruitment and a new finding of

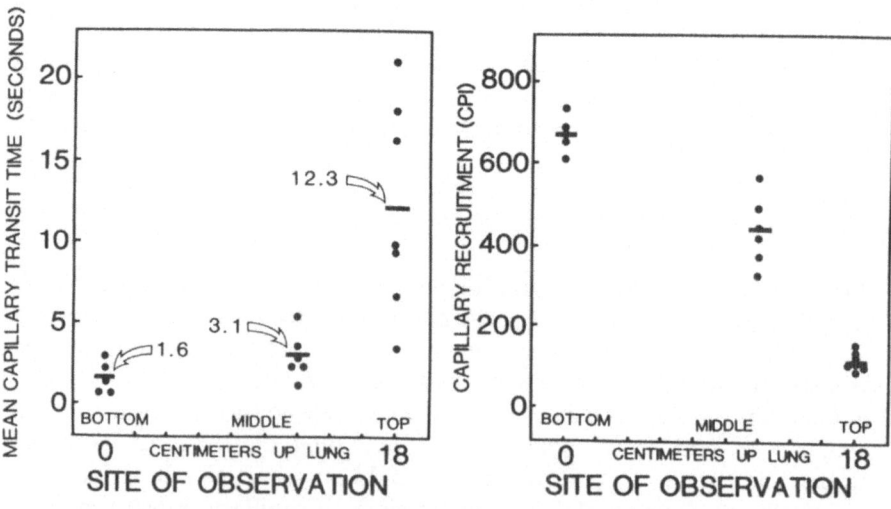

FIGURE 11.7. Mean capillary transit times (*left*) are rapid in the lower lung and slow in the upper lung. Capillary recruitment (*right*) reaches high levels in the lower lung, whereas not many capillaries are perfused by red cells in the upper lung. Observations were always made on the surface of the left lower lobe; each group of animals was positioned so that studies were made during different hydrostatic conditions (i.e., of the upper, middle, and lower lung relative to gravity). Data from the three groups are different from each other ($p < .025$). Scatter in upper lung data may indicate that pulmonary arterial pressure was so low that small changes resulted in large changes in transit times. *Bars* indicate means.

apparently equal importance to recruitment, that of a large potential for increasing red cell velocity.

In addition to being able to measure mean capillary transit time, it is possible to deconvolve the arteriolar and venular curves to obtain the capillary transport function. This function contains information about the transit through individual capillary networks. The range of transit times of the entire dye bolus can be calculated to obtain the distribution of transit times. These data should provide new information about a number of potentially interesting questions. For example, how homogeneously are capillary networks perfused between lung zones? Does perfusion homogeneity change with altered hemodynamics in fully recruited capillary beds? More difficult to predict is the particularly interesting question in the upper lung, where there are recruitable networks: How do hemodynamic alterations affect perfusion homogeneity?

References

Capen RL (1979), Latham LP, Wagner WW Jr (1981) Diffusing capacity of the lung during hypoxia: The role of capillary recruitment. *J Appl Physiol* 50:165–171.

Capen RL, Latham LP, Wagner WW Jr (1987) Comparison of direct and indirect measurements of pulmonary capillary transit times. *J Appl Physiol* 62:1150–1154.

Capen RL, Wagner WW Jr (1982) Intrapulmonary blood flow redistribution during hypoxia increases gas exchange surface area. *J Appl Physiol* 52:1575–1581.

Conhaim RL, Staub NC (1980) Reflection spectrophotometric measurements of O_2 uptake in pulmonary arterioles of cats. *J Appl Physiol* 48:848–856.

Dawson A, Grover RF (1974) Regional lung function in natives and long-term residents at 3,100 m altitude. *J Appl Physiol* 36:294–298.

Dawson CA, Grimm DJ, Linehan JH (1979) Lung inflation and longitudinal distribution of pulmonary vascular resistance during hypoxia. *J Appl Physiol* 47:532–536.

de Alva WE, Rainer J, Filley GF (1962) Cinemicrography of the living lung. *Anat Rec* 142:349.

Dugard A, Naimark A (1967) Effect of hypoxia on distribution of pulmonary blood flow. *J Appl Physiol* 23:663–671.

Grover RF (1963) *Normal and Abnormal Pulmonary Circulation.* Karger, Basel.

Hales S (1733) *Statistical Essays: Containing Haemastaticks.* (Reprinted 1964.) Hafner, New York.

Jameson AG (1964) Gaseous diffusion from alveoli into pulmonary arteries. *J Appl Physiol* 19:488.

Kato M, Staub NC (1966) Response of small pulmonary arteries to unilobar hypoxia and hypercapnia. *Circ Res* 19:426–439.

Krahl VE (1963) A method of studying the living lung in a closed thorax, and some preliminary observations. *Angiology* 14:149–159.

Malpighi M (1961) De Pulmonibus. (Transl by Young J, 1929.) *Proc Royal Soc Med* 23:7.

Nagasaka Y, Bhattacharya J, Nanjo S, Gropper MA, Staub NC (1984) Micropuncture measurement of lung microvascular pressure profile during hypoxia in cats. *Circ Res* 54:90–95.

Permutt S, Bromberger-Barnea B, Bane HN (1962) Alveolar pressure, pulmonary venous pressure, and the vascular waterfall. *Med Thorac* 19:239–260.

Sackner MA, Will DH, Dubois AB (1966) The site of pulmonary vascular activity during hypoxia and serotonin administration. *J Clin Invest* 45:112–120.

Shirai M, Sada K, Ninomiya I (1986) Effects of regional hypoxia and hypercapnia on small pulmonary vessels in cats. *J Appl Physiol* 16:40–448.

Staub NC (1961) Gas exchange vessels in cat lung. *Fed Proc* 20:107.

Wagner PD (1977) Diffusion and chemical reaction in pulmonary gas exchange. *Physiol Rev* 57:257–312.

Wagner WW Jr (1969) Pulmonary microcirculatory observations in vivo under physiological conditions. *J Appl Physiol* 26:375–377.

Wagner WW Jr, Barker DB, Filley GF (1967) A photographic method for quantitating blood flow in the pulmonary microcirculation. *J Biol Photo Assoc* 35:95–108.

Wagner WW Jr, Brinkman PD, Barker DB, Filley GF (1969) Erythrocyte photomicrography: Contrast control by monochromatic transillumination. *J Biol Photo Assoc* 37:156–164.

Wagner WW Jr, Filley GF (1965) Microscopic observation of the lung in vivo. *Vasc Dis* 2:229—241.

Wagner WW Jr, Latham LP (1975) Airway hypoxia causes pulmonary capillary recruitment in the dog *J Appl Physiol* 39:900–905.

Wagner WW Jr, Latham LP, Capen RL (1979) Capillary recruitment during airway hypoxia: The role of pulmonary artery pressure. *J Appl Physiol* 47:383–387.

Wagner WW Jr, Latham LP, Gillespie MN, Guenther J, Capen RL (1982) Red cell transit times across pulmonary capillaries. *Science* 218:379–380.

Wagner WW Jr, Latham LP, Hanson WL, Hofmeister SE, Capen RL (1986) Vertical gradient of pulmonary capillary transit times. *J Appl Physiol* 61:1270–1274.

Wearn JT, Ernestene AC, Bromer AW, Barr JS, German WJ, Zschiesche LJ (1934) The normal behavior of the pulmonary blood vessels with observations on the intermittence of the flow of blood in arterioles and capillaries. *A J Physiol* 109:236–256.

West JB (1979) *Respiratory Physiology.* Williams & Wilkins, Baltimore.

12
Elasticity of Microvessels in Postmortem Human Lungs

RONG-TSU YEN

Introduction

Recently we have demonstrated that in the case of cat, after the necessary morphometric and rheological data are collected, we can predict the pulmonary blood flow as a function of a number of physiological parameters with excellent accuracy. We can use the computing program to assess in vivo parameters such as the pulmonary surface (Engelberg et al., 1959; Yen and Foppiano, 1981; Yen et al., 1980; Zhuang et al., 1985). We believe that by analogy we can do the same for the human. Such a valid biophysical mathematical analysis of the pulmonary circulation of humans may have considerable value in further understanding the normal pulmonary circulation. An attempt is made to obtain a complete set of morphological and rheological data on the human lung from postmortem material obtained within 24 hours of death, in order to initiate and establish a theoretical model and computing program of pulmonary circulation. The model will be validated as far as possible from experimental data derived from flow experiments on isolated postmortem human lungs. It will correlate many physiological variables through a rational theory: the pressure in pulmonary artery, left atrium, airway, and pleura; the branching pattern of the vascular tree, including the branch number ratio, the diameter ratio, and the length ratio of all orders of vessels; the alveolar surface area; the capillary blood vessel characteristics; the rheological properties, including the distensibilities of all generations of blood vessels; and the flow properties of blood. The model and computing program can be used to help interpret clinical data and promote clinical research. By fitting certain clinical data into the computing program, several morphological and rheological parameters of the patient can be determined. The values of these parameters should be useful for diagnosis or evaluation of treatment modalities by clinicians in the future and thus making a big step forward in human pulmonary physiology.

Earlier the cat was used not as a model of man, but as an example of our mathematical approach. It is generally believed that a perfect animal model of human pulmonary circulation is not known. In fact, the literature contains many examples showing significant differences between man and dog, or cat, or rabbit. For example, Kalk et al. (1975) showed that the equidiameter curves for human and dog pulmonary arteries are very different. Figure 12.1 shows the values of the pulmonary artery pressure (P_a) and the transpulmonary pressure ($P_{tp} = P_a - P_{PL}$; P_{PL} is pleural pressure) at which the arterial diameter remained constant. Examples like this tell us that morphological and rheological data on the human are of great relevance to human physiology. One cannot rely on animal data.

Since there is no way to obtain a complete set of human data on living persons, the use of postmortem material is the logical answer. In our experiments, all lungs obtained at autopsy are from individuals with sudden, non-hospital death. Lungs heavier than 475 g frequently have microscopic evidence of alveolar edema and are not accepted for study. Similarly, lungs with evidence of hemorrhage, excessive tracheal secrections, or large pleural tears are excluded. The study is restricted to the left lung and is completed within 24 hours after death. The experimental protocol is initiated as soon as the specimen is obtained, and the lungs are kept at 4°C until the experiment is started.

At the present time, morphological data on human airways, pulmonary arteries, and veins are known due to work by Weibel (1963), Singhal et al. (1973), and Horsfield and Gordon (1978, 1981). The pulmonary capillary blood vessel structure of human alveoli is not known. Data on the distensibility of human pulmonary blood vessels do not exist. Of nonhuman

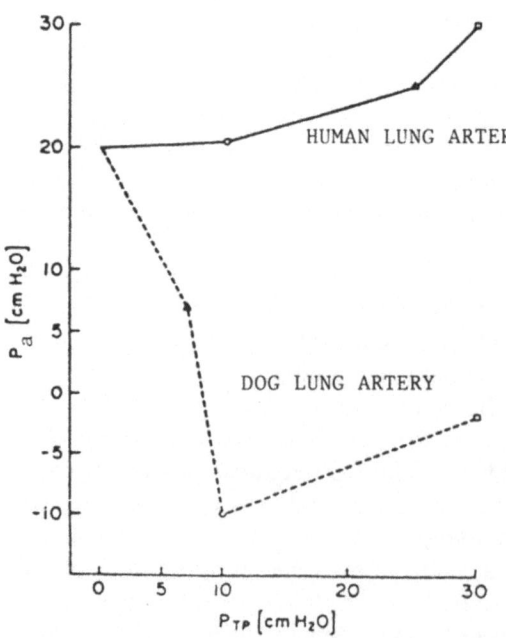

FIGURE 12.1. Comparisons of equidiameter curves for human and dog pulmonary arteries. P_a = pulmonary artery pressure; P_{TP} = transpulmonary pressure. (From Kalk et al., 1975.)

mammals, a full set of morphological and rheological data has been obtained for the cat by the author (Yen and Foppiano, 1981; Yen and Fung, 1973; Yen et al., 1980, 1983, 1984). For the dog, there exists a wealth of physiological data, but the dog's morphological and rheological data are missing and inappropriate for human correlation. On the other hand, a large amount of clinical data on the human lung exists (Fishman and Hecht, 1969). If interpretation of these data can be more definitive and their use can be made more reliable, much more data can be obtained by clinicians in the future. This human lung study could open up a vast frontier for future pulmonary circulation research, with direct relevance to man. Reported here are our studies of the pressure-diameter relationships of the postmortem human lung in the noncapillary segments of arterial and venous portions of the pulmonary microcirculation by the silicone elastomer perfusion techniques used by us in previous studies (Sobin et al., 1972).

Methods

Silicone Elastomer Preparation

We measured the diameters and branching hierarchy of various segments of the pulmonary microvasculature from the histological preparations of postmortem human lung prepared by perfusion with a silicone elastomer with methods previously described in detail (Sobin et al., 1972). The noncatalyzed fluid silicone elastomer (General Electric component 88017) readily passes the pulmonary capillary bed. Distinguishing between pulmonary arterial and venous vessels, is accomplished by first perfusing with the clear catalyzed material the entire vasculature, including the capillary bed, and then using the same material containing 0.25% Cab-O-Sil (Cabot Corporation, Boston, MA) to fill and obstruct either the pulmonary arterial-arteriolar vessels (antegrade perfusion) or the venous-venular vessels (retrograde perfusion). The initial viscosity of 17–20 cp was not significantly changed for the first 20 min after addition of the catalytic materials ethyl silicate (to 1.5%) and tin octate (to 5.5%), but subsequently viscosity increased steadily with firm gelation by 2 h.

The studies were carried out on three postmortem lungs (Table 12.1). In all cases the cause of death was accidental and did not involve the lung. The lung collapsed with thoracotomy, which indicated detained elasticity and a clear

TABLE 12.1.   Specimen information

Code	Age (yr)	Sex	Height (in)	Weight (lb)	Hours P.M.	Preparation	Lung weight (g)	Perfusion direction
CUZ	30	F	65	132	30	Right lung	275	Antegrade
CUO	28	M	70	154	7	Left lung	175	Retrograde
CUP	30	M	64	160	8	Left lung	300	Antegrade

airway. The lungs were removed by transection of the trachea 5 cm above the carina, and the trachea was cannulated after gentle suction to remove retained secretions. Only half-lungs were used in the experiment and the other halves were retained for histopathology. In the experiment, the pulmonary artery and left atrium were cannulated. The lung was initially inflated to 20 cm H_2O and cyclically inflated from 4 to 14 cm H_2O with periodic inflation to 20 cm H_2O to remove areas of superficial atelectasis. A small amount of saline was perfused from the pulmonary artery or vein to the opposing vein or artery, respectively, to establish vascular continuity across the lung. This was followed by removal of saline from the tubing, and then perfusion of the catalyzed silicone elastomer was begun. After silicone was visible in the exit line the perfusion container was promptly replaced by silicone-containing Cab-O-Sil, catalyzed as it was prepared. The level of silicone infusion was initially set at 34 cm H_2O above the base of the vertically placed lung and was maintained at that level during infusion. This established a gradient of vascular pressure in the pulmonary vessels with a maximum vascular transmural pressure of 24 cm H_2O. As silicone fluid level dropped during the first moments of infusion the level of the infusion bottle was raised to maintain initial pressure. Three hours after addition of the catalyst to the liquid silicone, the vascular tubing connections were cut and the tracheal cannula was attached to a container of 3.7% buffered formaldehyde solution. The lung was instilled with formaldehyde solution that continually seeped through the pleura. The level of the fixative was again set 27 cm above the lung base and so maintained by pump-recycling the formaldehyde for 48 hours.

Histological Preparation

Blocks of tissue were removed from the lung, identified as to their position, and marked to indicate top and bottom of each block. Selected samples of blocks were embedded in 10% gelatin solution, and after the gelatin hardened in formaldehyde solution, sectioned at 200 μm, and mounted in glycerol without staining. The tissue block hydrostatic level from which slides were prepared was carefully noted on each side. The histoligical slides were scanned on an Ortholux microscope under low magnification, and areas were taken for selection of the noncapillary vessels to be measured and hierarchy of that branch determined. Vessel diameter was measured under a 40 × objective magnification by a dimension analyzer, using an electronic image-splitting or shearing device.

Results

Figures 12.2 and 12.3 are photomicrographs of the postmortem lung perfused with silicone elastomer containing colloidal silica that prevented filling of the capillary bed proper. Figure 12.2 is of a precapillary arteriole, showing filling

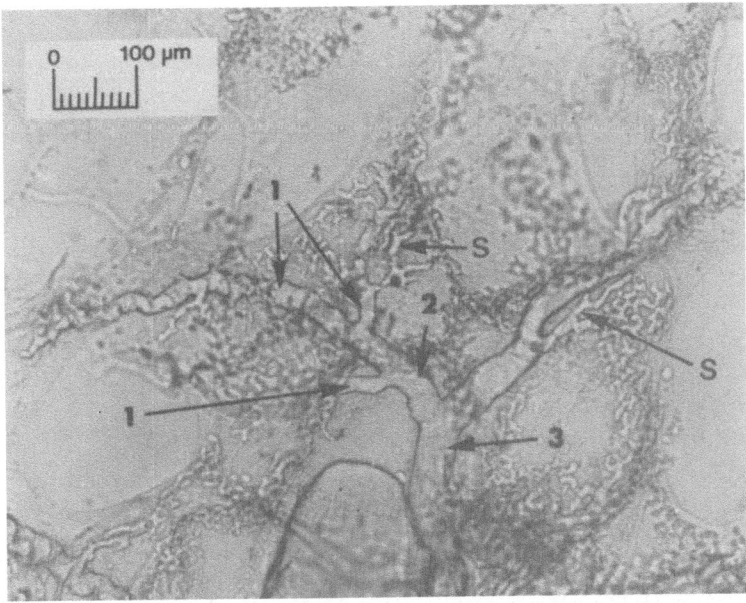

FIGURE 12.2. Typical photomicrograph of postmortem lung perfused with silicone elastomer containing silica that prevented filling of capillary bed proper. This figure shows several orders of precapillary arterioles. Smallest noncapillary vessels are identified and designated as order 1 and junction of nonfilled capillary sheet is indicated by "S". Transmural pressure (local blood vessel pressure minus airway pressure) is 16.7 H_2O in this case.

of the capillary bed. The silica-filled mass is highly refractable, and it is sharply differentiated from the transparent adjacent capillaries of the sheet (Fig. 12.2 and 12.3). With the use of the Strahler system for hierarchy classification (Strahler, 1957) the smallest noncapillary vessel is identified. The junction of the capillary sheet is indicated by "S" and the smallest noncapillary vessel is designated as order 1. Two order 1 vessels converge as order 2, followed by order 3, the next in this ordering series. This branching order applies to both the pulmonary artery system (Fig. 12.2, antegrade perfusion) and pulmonary venous system (Fig. 12.3, retrograde perfusion).

The results of measurement for pulmonary arterioles are given in Figure 12.4 and pulmonary venules in Figure 12.5, in which vessel diameter is plotted against transmural pressure and where lung inflation is 10 cm H_2O. Table 12.2 summarizes these data for both arterioles and venules. D_0 gives the diameter of each group of vessels at zero pressure. The vessel compliance coefficient α is computed as the slope of the regression lines (Figs. 12.3 and 12.4) and is expressed as microns per centimeter of water incremental pressure. The compliance coefficient is also expressed as the percentage change per

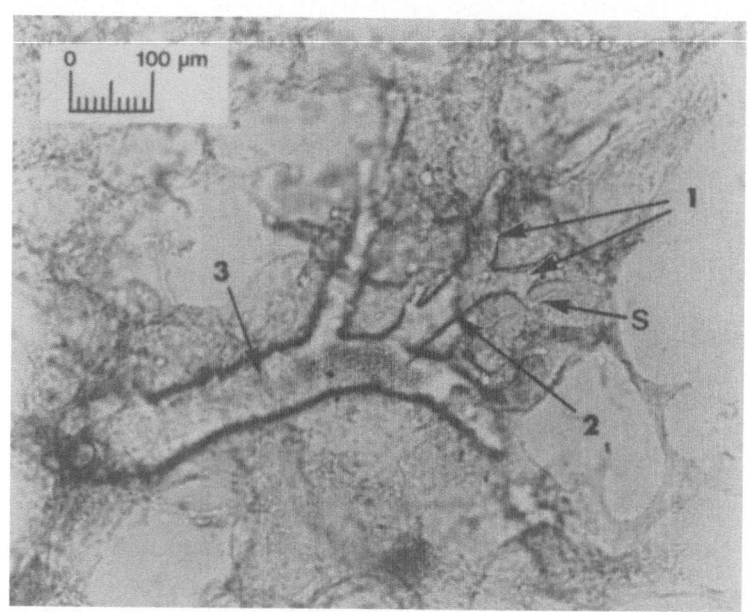

FIGURE 12.3. Typical photomicrograph of postcapillary bed venule. Transmural pressure is 12.4 cm H_2O for this particular figure. Please note the difference of transmural pressure between Figures 12.2 and 12.3. That is why vessel diameter of the venules appears to be smaller than that of the arterioles. S, junction of nonfilled capillary sheet.

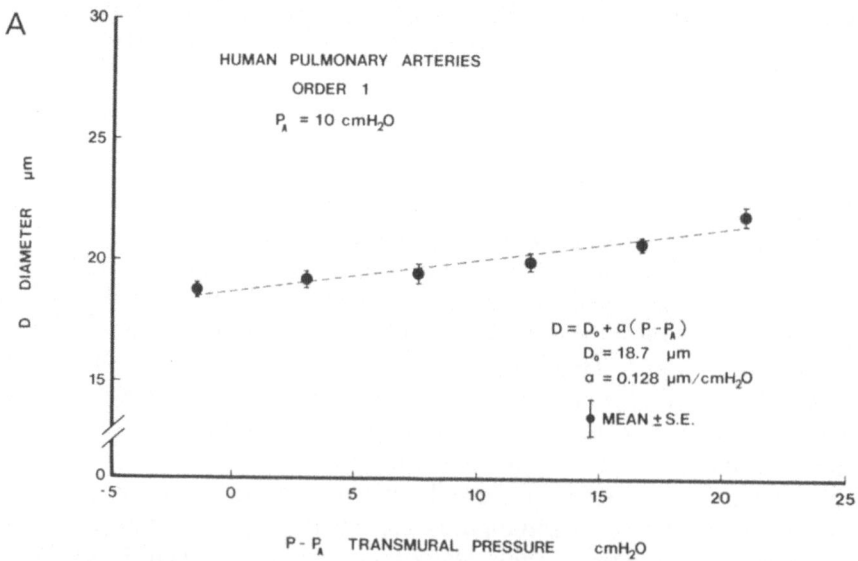

FIGURE 12.4. Vessel diameter is plotted against transmural pressure for arterioles. P_A, alveolar pressure; P, pressure; D_0, value of diameter when transmural pressure is at 0; D, diameter; α, vessel compliance coefficient. A: Arteriole of order 1. An average of 158 measurements were made at each point. $r = .97$. B: Arteriole of order 2. An average of 187 measurements were made at each point. $r = .97$. C: Arteriole of order 3. An average of 101 measurements were made at each point. $r = .97$.

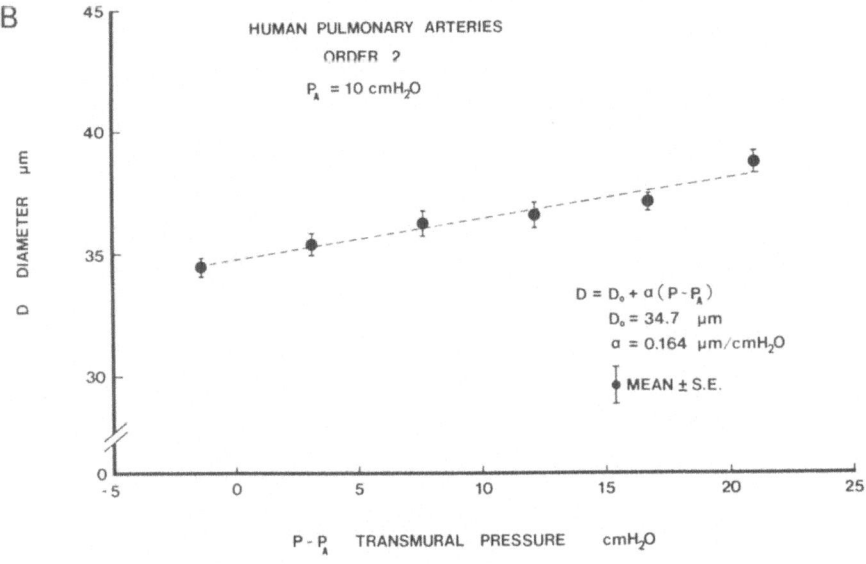

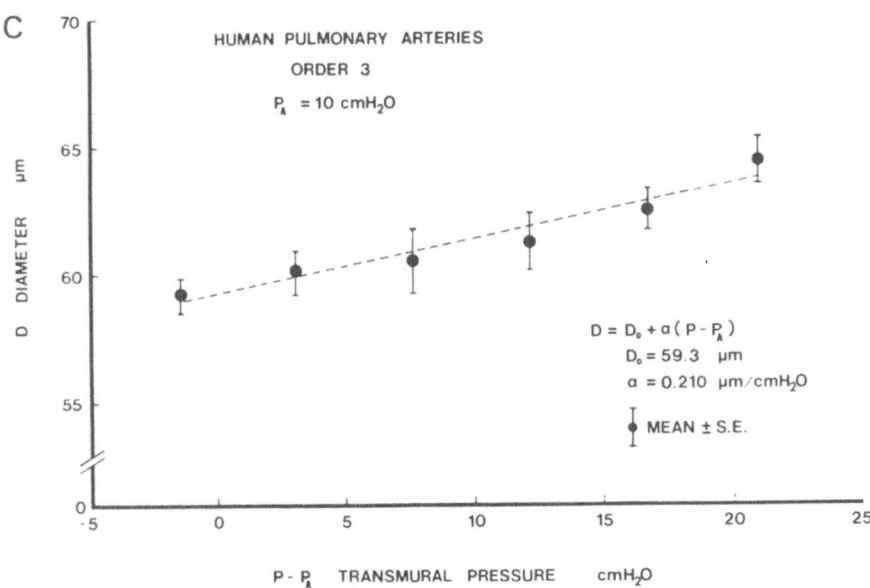

FIGURE 12.4 *Continued*

centimeter of water, normalized to D_0 for the purpose of comparison to other forms of compliance. The values of the compliance coefficient are 0.128, 0.164, and 0.210 μm/cm H_2O or 0.682, 0.472, and 0.354%/cm H_2O, respectively, of orders 1 through 3 for arterioles. The compliance coefficient for orders 1 through 3 venules is 0.187, 0.215, and 0.250 μm/cm H_2O or 0.992, 0.612, and 0.424%/cm H_2O, respectively. For these microvessels, the smaller vessels tend

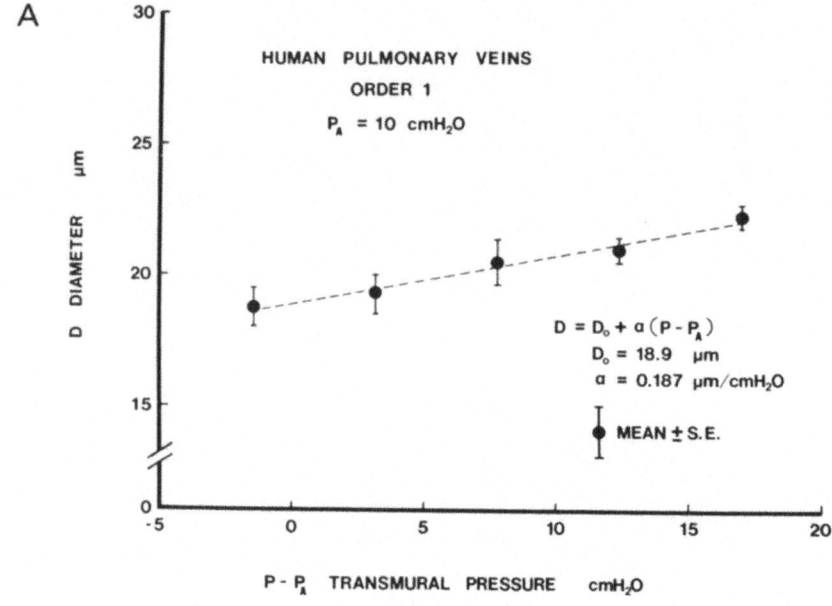

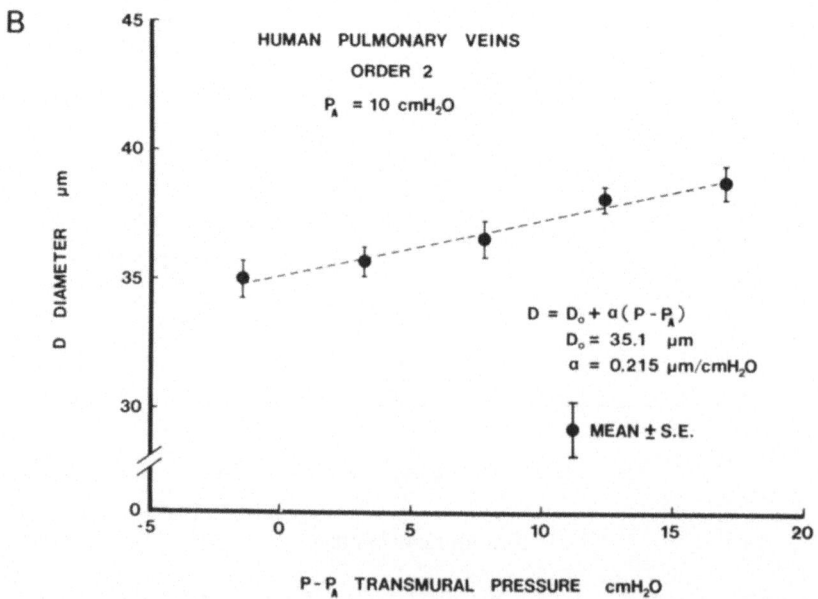

FIGURE 12.5. Relationship between vessel diameter and transmural pressure for venules. *A*: Venules of order 1. An average of 43 measurements were made at each point. $r = .99$. *B*: Venules of order 2. An average of 49 measurements were made at each point. $r = .99$. *C*: Venules of order 3. An average of 30 measurements were made at each point. $r = .99$. Abbreviations as for Figure 12.4.

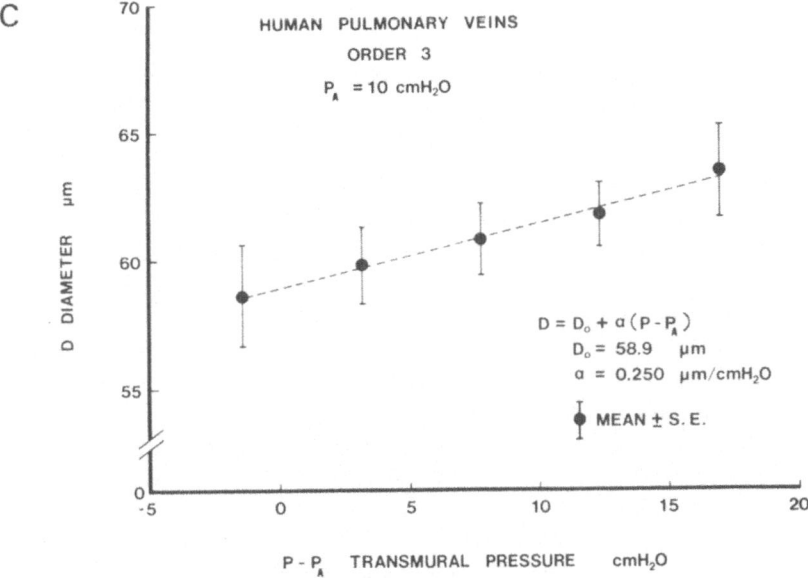

FIGURE 12.5 *Continued*

TABLE 12.2.   Elasticity of small noncapillary pulmonary blood vessels of the human lung*

Order	Transpulmonary pressure $(P_A - P_{PL}) = 10$ cm H_2O	
	D_0 (μm)	α (μm/cm H_2O)
Arterioles		
3	59.3 $\pm$ 7.1	0.210
2	34.7 $\pm$ 5.8	0.164
1	18.7 $\pm$ 3.9	0.128
Venules		
1	18.9 $\pm$ 3.0	0.187
2	35.1 $\pm$ 3.5	0.215
3	58.9 $\pm$ 7.9	0.250

* Values are means $\pm$ SD. α, Compliance coefficient; D_0, vessel diameter at 0 transmural pressure; P_A, alveolar pressure; P_{PL}, pleural pressure.

to be more compliant than the larger ones. The venules are somewhat more compliant than the corresponding order of arterioles. As for the significance of the diameter versus transmural pressure relationship, Student's *t* test shows that the significance level is 99.9, 99, and 99.9% for arterioles of the order 1 through 3, respectively, and 99.5, 95, and 97.5% for venules of the orders 1 through 3, respectively.

Discussion

Patency of Postmortem Pulmonary Microcirculation

It is important to note that the vascular bed of the lung is made relatively agonally (i.e., at the time of death). Miller (1974), Von Hayek (1960), and Harris and Heath (1962), among others, have filled the capillary bed with perfused gelatin and other materials, and Weibel's (1963) elegant morphometric studies on the postmortem lung required that the microvasculature was bloodless. Sobin et al. (1979) perfused the postmortem human lung with blood from the pulmonary artery as well as from the pulmonary veins, established a gradient of pressure down the pulmonary vasculature, and quick froze a linear band down the lung. Tissues from such lungs showed a linear microvascular sheet dimension change in accord with the pressure gradient, indicating a capillary sheet compliance of 0.127 μm/cm H_2O over a range of 5–30 cm H_2O. Such observations strongly support the patency of the microcirculation in the postmortem human lung.

The reason why the vascular bed is relatively empty agonally is probably similar to that for the heart, which made the Greeks miss the opportunity for discovering circulation. Fahraeus (1975) explained it by the dilation of small arteries after death. He measured the pressure in the carotid and femoral arteries in corpses 24 hours after death and found them to be negative, varying from a few millimeters to 1 cm Hg. In my work, I found no difficulty perfusing postmortem lungs of the human.

Methodology

The silicone microvascular casting methods developed by us are suited to determine branching hierarchy and vascular diameter of branches of the microcirculation. Validation of their use has been discussed in our prior publications (Fung et al., 1983; Sobin et al., 1972, 1980). The silicone elastomer has the following desirable characteristics for microvascular casting: initial low viscosity, catalytic nonexothermic controllable rate of polymerization, dimensional stability on catalysis to a firm elastomer, nonpolar behavior and thus nonmiscibility with blood, and nontoxicity (acutely) to the endothelium. The miscibility of the silicone with colloidal silica forms masses that do not enter the capillary bed and thus, like a thrombus, allow selective filling of either arterial and venous side of the microcirculation. This combination of properties led to our investigating the end three branches of the pulmonary arterial and venous microcirculation.

Microvessel Elasticity

Although the airways have been studied extensively in the postmortem human lung, the vascular system has been less commonly studied. Singhal and co-workers (1973) and Horsfield (1978) investigated the morphometry of the

TABLE 12.3.   Comparisons of our results with the results of Horsfield and Gordon (1981)*

| | Transpulmonary pressure $(P_A - P_{PL}) = 10$ cm H_2O | | | |
| | D_0 (μm) | | Diameter ratio | |
Order	This report	Horsfield and Gordon (1981)	This report	Horsfield and Gordon (1981)
	Arterioles			
1	18.7	13		
2	34.7	21	0.51	0.62
3	59.3	34	0.59	0.62
	Venules			
1	18.9	13		
2	35.1	19	0.54	0.68
3	58.9	29	0.59	0.68

* D_0, vessel diameter at 0 transmural pressure; diameter ratio, ratio of diameters in successive orders; P_A, alveolar pressure; P_{PL}, pleural pressure. Results of Horsfield and Gordon are obtained from their papers in 1978 and 1981.

postmortem human pulmonary arteries; the former studies described arterial branching at the proximal end of microcirculation (arteries 0.138 mm in diameter) and the latter studies described branching pattern in the immediate four orders proximal to the alveolar capillaries. Horsfield and Gordon (1981) then studied casts of pulmonary veins in postmortem human lungs, determined branching and dimension hierarchy out to vessels of 200 μm diameter, and synthesized a full vascular model comparing pulmonary arterial and venous branches.

Table 12.3 compares our measured dimension of orders 1 through 3 for pulmonary arterioles and venules, and ratio of a branch with a next order, with the same parameters of Horsfield and Gordon (1981). There is a reasonable consistency between our data and the data of Horsfield and Gordon, including the ratio of vessel dimensions in successive orders. Although order 1 vessels are usually the first or initial branch in a network, we have excluded the capillary bed in our numbering system, and our Figures 12.2 and 12.3 are clearly so labeled. Unfortunately, the description of Horsfield and Gordon is less precise. Their order 1 branches are defined as the first branch of 10–15 μm in diameter down any pathway. Thus their order 2 branch is equivalent to our order 1.

Comparisons of Microvessel Elasticity Between Human and Cat Lung

Table 12.4 shows the comparisons of the microvessel elasticity between human and cat lung. In the cat juxta-alveolar vessels of less than 50 μm Sobin et al. (1978) noted a compliance of 0.274 μm/cm H_2O. This is approximately double

TABLE 12.4. Comparison of elasticity of small pulmonary blood vessels between human and cat lung*

Blood vessels	Order	Human D_0 (μm)	Human α (μm/cm H_2O)	Cat D_0 (μm)	Cat α (μm/cm H_2O)
		Transpulmonary pressure $(P_A - P_{PL}) = 10$ cm H_2O			
Arterioles	3	59.3	0.210	73	0.274
	2	34.7	0.164	44	
	1	18.7	0.128	24	
Capillary Sheet	h_0	3.5	0.127	4.28	0.219
Venules	1	18.9	0.187	25	0.274
	2	35.1	0.215	46	
	3	58.9	0.250	77	

* α, compliance coefficient; D_0, vessel diameter at 0 transmural pressure; h_0, sheet thickness at 0 transmural pressure; P_A, alveolar pressure; P_{PL}, pleurel pressure.

the compliance coefficient of the microvessels in orders 1 through 3 in the postmortem human lung. Similarly, the compliance of the cat alveolar capillaries is 0.217 μm/cm H_2O (Sobin et al., 1972) and is approximately twice that of the human alveolar capillaries. However, in both the human and the cat, there is a consistency of elasticity between the alveolar microvascular sheet and orders 1 through 3 or juxta-alveolar vessels. This internal consistency is understandable, and the difference between cat and human implies a structural difference between the two species. The characteristic paucity of smooth muscle in these arterioles and venules is seen in both human (Wagenvoort et al., 1964) and cat (Sobin et al., 1980) lung. Thus the elastic behavior of both the capillary bed and the juxta-alveolar vessels appears to be determined by the properties of an endothelium-lined tube with its basal lamina, by fibrils of collagen and occasional fibroblasts, and by minimal but importantly placed vascular smooth muscle.

Validity of Measurements Obtained from Postmortem Material

Evidence from Electron Microscopy

Bachofen and coworkers (1975) found no changes in the postmortem human lung by transmission electron microscopy up to 8 h postmortem. The significance of these observations is increased when it is considered that the lung remains in the body for that period of time. Such observations place increased importance on the use of such tissues and organs.

Elasticity

Blood vessel elasticity is measured by the pressure-diameter relationship. Data on the elasticity of human and nonhuman (mammalian) blood vessels have been almost completely obtained from postmortem tissues (Bader, 1963; Dobrin, 1984; McDonald, 1960; Remington, 1957). The literature is extensive; it includes vessels from the aorta down to the microcirculation and large vein capacitance. Bader (1963) illustrated "stretch curves" of a pig aorta made shortly after death and 8 days later; the curves are essentially duplicative (i.e., there was no significant difference). McDonald (1960) discussed "the comparison of the isolated specimen and its behavior in the body" and stated categorically that misconceptions arise from removal of an artery from the body. In discussing the validity of postmortem tissues for physiological study he noted that there is no evidence that the physical properties of collagen and elastin alter with blood supply. With the constraints of studying the elasticity of the pulmonary vessels in situ in the intact lung, there is no evidence that vascular elasticity as measured is other than valid.

Sugihara et al. (1971) studied the length-tension relationship of alveolar wall. They used 170 pieces of lung parenchyma removed from 36 individuals at surgery or autopsy. All experiments were performed at room temperature and were completed within 10 h of obtaining the material. Some autopsy material was delayed in its delivery to their lab as long as 36 h after death. They consider these data are valid based on the studies of Fukaya et al. (1968) to measure the mechanical properties of alveolar walls of the cat. Length-tension experiments on the lung tissues, which were removed from normal cats 0.1–15 h after death, were performed 30 min after death and at repeated intervals for 36 h. Over this period, the length-tension relationships of the tissue did not change appreciably.

Biophysical-Physiological Basis of the Study

Although physiological experiments are consonant with the living state, such experiments can be simulated in the postmortem lung and could therefore be considered to be mechanically based. Certainly there is nothing in this procedure that implies that the lung is "living." The information obtained relates to the interrelationships among the various mechanical factors influencing blood flow through the lung. It is noted initially that valid morphometric data required open communication through the pulmonary capillary bed and that the casting experiments of Horsfield et al. and ours and the blood perfusion of the postmortem human lung by Sobin et al. met this provision.

Validity of the Assumption That the Vessels of a Given Order Have Identical Properties Irrespective of Their Base-to-Apex Location

In the microvascular elasticity, it is not possible to observe a single vessel and follow its change in diamter without changing transmural pressure. The best one can do is to identify the hierarchy of the vascular tree and measure the mean diameter as a function of changing pressure difference. In doing so, we assumed that the microvessels of a given order have identical physical proper-

TABLE 12.5.   Human pulmonary arteries measured in different lobes under a constant transmural pressure*

Order	n	Upper lobe diameter (μm)	n	Middle lobe diameter (μm)	n	Lower lobe diameter (μm)
4	52	86.5 ± 6.9	80	87.8 ± 8.3	57	87.6 ± 7.7
5	140	134.5 ± 9.8	187	136.2 ± 10.1	181	137.1 ± 10.2
6	148	213.8 ± 14.4	170	216.1 ± 14.2	165	216.4 ± 14.9
7	102	330.6 ± 21.8	135	334.8 ± 25.7	124	333.8 ± 21.9
8	71	529.5 ± 37.2	71	521.3 ± 34.2	93	517.9 ± 34.9
9	38	816.8 ± 58.6	35	815.3 ± 61.5	41	785.0 ± 69.4
10	16	1195.1 ± 109.2	17	1235.5 ± 160.1	19	1130.6 ± 110.2
11	5	1785.0 ± 134.5	5	1717.9 ± 145.8	6	1673.0 ± 153.2

* Values are means ± SD; n is number of vessels measured. Human left lung: age, 51 yr; sex, male; weight, 195 lb; height, 75 in; alveolar pressure, 10 cm H_2O; pleural pressure, 0; transmural pressure (local blood pressure − alveolar pressure) = 15 ± 1 cm H_2O.

ties irrespective of their base-to-apex location. To address the validity of this assumption, we performed the silicone elastomer casting experiments on the pulmonary arterial tree with the lung in a horizontal position. We measured the dimension of the vessels from bared casts, sampling the lung at different points from apex to base. These were at identical vertical positions to ensure identical transmural pressures. The results are listed in Table 12.5. It shows that for the vessels up to order 8, the same-order artery at the lower lobe or middle lobe has the same diameter as the same-order artery at the apex when the vessels are exposed to the same transmural pressure (the transmural pressure is 15 ± 1 cm H_2O in this case). When the vessel is higher than order 8, this vertical uniformity is less evident, and a different method should be applied in measuring the vessel elasticity [i.e., x-ray photography method (Caro and Saffman 1965; Yen and Foppiano, 1981; Yen et al., 1980)].

Postmortem Data Must Be and Can Be Properly Used— The Potential Use of These Data

The postmortem data must be used properly. With proper use, an error analysis can be made and tolerable limits of errors in individual quantities can be computed. Through such analyses one can learn to use the data intelligently.

In our work, we shall use the postmortem morphological and rheological data to establish a theoretical computing program to relate the blood flow with the pressures, geometry, and rheological parameters. After the program is validated, some morphological and rheological parameters may be freed up and treated as unknowns. Clinical data can then be used to determine these parameters by others.

References

Bachofen M, Weibel ER, Roos B (1975) Postmortem fixation of human lungs for electron microscopy. *Am Rev Resp Dis* 111:247–256.

Bader H (1963) The anatomy and physiology of the vascular wall. In *Handbook of Physiology, Sec 2, Circulation*, Vol 2. American Physiological Society, Washington, DC, pp 865–889.

Caro CG, Saffman PG (1965) Extensibility of blood vessels in isolated rabbit lungs. *J Physiol London* 178:193–210.

Dobrin PB (1984) Biomechanics of arteries and veins; mechanical properties. In Abramson DI, Dobrin PB (eds) *Blood Vessel and Lymphatics in Organ System*. Academic, New York.

Engelberg, J, DuBois AB (1959) Mechanics of pulmonary circulation in isolated rabbit lungs. *Am J Physiol* 196:401–414.

Fahraeus R (1975) Empty arteries. Lecture delivered at the 15th International Congress of the History of Medicine, Madrid.

Fishman AP, Hecht HH (eds) (1969) *The Pulmonary Circulation and Interstitial Space*. University of Chicago Press, Chicago.

Fukaya H, Martin CJ, Young AC, Katsura S (1968) Mechanical properties of alveolar walls. *J Appl Physiol* 25:689–695.

Fung YC, Sobin SS, Tremer H, Yen MRT, Ho HH (1983) Patency and compliance of pulmonary veins when airway pressure exceeds blood pressure. *J Appl Physiol* 54:1538–1549.

Harris P, Heath D (1962) *Human Pulmonary Circulation*. Williams & Wilkins, Baltimore.

Horsfield K (1978) Morphometry of the small pulmonary arteries in man. *Circ Res* 42:593–597.

Horsfield K, Gordon WI (1981) Morphology of pulmonary veins in man. *Lung* 159:211–218.

Kalk J, Benjamin J, Comite H, Hutchins G, Traystman R, Menkes H (1975) Vescular interdependence in postmortem human lungs. *Am Rev Resp Dis* 112:505–511.

McDonald DA (1960) *Blood Flow in Arteries*. Edward Arnold, London.

Miller WS (1974) *The Lung* (2nd ed). Charles C Thomas, Springfield, IL.

Remington JW (ed) (1957) *Tissue Elasticity*. American Physiological Society, Washington, DC.

Singhal S, Henderson R, Horsfield K, Harding K, Cumming G (1973) Morphometry of the human pulmonary arterial tree. *Circ Res* 33:190–197.

Sobin SS, Fung YC, Lindal RG, Tremer HM, Clark L (1980) Topology of pulmonary arterioles, capillaries and venules in the cat. *Microvasc Res* 19:217–233.

Sobin SS, Fung YC, Tremer HM, Rosenquist TM (1972) Elasticity of the pulmonary alveolar microvascular sheet in the cat. *Circ Res* 30:440–450.

Sobin SS, Lindal RG, Fung, YC, Tremer HM (1978) Elasticity of the smallest non-capillary pulmonary blood vessels in the cat. *Microvasc Res* 15:57–68.

Sobin SS, Tremer HM, Lindal RG, Fung YC (1979) Distensibility of human pulmonary capillary blood vessels in the interalveolar septa (abstract). *Fed Proc* 38:990.

Strahler AN (1957) Quantitative analysis of watershed geomorphology. *Trans Am Geophys Union* 38:913–920.

Sugihara T, Martin CJ, Hildebrandt J (1971) Length-tension properties of alveolar wall in man. *J Appl Physiol* 30:874–878.

Von Hayek H (1960) *The Human Lung.* (English translation, revised and edited by VE Krahl.) Hafner, New York.

Wagenvoort CA, Heath D, Edwards JE (1964) *Pathology of Pulmonary Vasculature.* Charles C Thomas, Springfield, IL.

Weibel ER (1963) *Morphometry of the Human Lung.* Academic Press, New York.

Yen RT, Foppiano L (1981) Elasticity of small pulmonary veins in the cat. *J Biomech Eng Trans ASME* 103:38–42.

Yen RT, Fung YC (1973) Model experiments on apparent blood viscosity and hematocrit in pulmonary alveoli. *J Appl Physiol* 35:510–517.

Yen RT, Fung YC, Bingham N (1980) Elasticity of small pulmonary arteries in the cat. *J Biomech Eng Trans ASME* 102:170–177.

Yen RT, Zhuang FY, Fung YC, Ho HH, Tremer H, Sobin SS (1983) Morphometry of the cat's pulmonary venous tree. *J Appl Physiol: Resp Environ Exercise Physiol* 55:236–242.

Yen RT, Zhuang FY, Fung YC, Ho HH, Tremer H, Sobin SS (1984) Morphometry of the cat's pulmonary arteries. *J Biomech Eng* 106:131–136.

Zhuang FY, Yen RT, Fung YC, Sobin SS (1985) How many pulmonary alveoli are supplied (drained) by an arteriole (venule)? *Microvasc Res* 29:18–31.

13
Connection of Micro- and Macromechanics of the Lung

Yuan-Cheng Fung

Introduction

In Chapter 1 Dr. Zweifach described the history of microcirculation research, and a perspective of the future. The following five chapters describe new findings about coronary and peripheral microcirculation. In Part II of this book, there are six chapters describing new assessments of pulmonary circulation. Why is it desirable to divide the field of microcirculation into these two halves? One reason is the existence of a basic difference in the mechanical properties of the pulmonary and peripheral capillary blood vessels. Capillaries of the peripheral organs, including the heart wall, can be roughly characterized as being "rigid." For these vessels, there is little measurable change of vessel diameter under an optical microscope when blood pressure changes within physiological range. Under negative transmural pressure the peripheral capillaries do not collapse. On the other hand, capillaries of the lung are very "distensible," and they are readily collapsible under negative transmural pressure. As a consequence, the capillaries of the peripheral organs play a different role in the pressure-flow relationship than those in the lung. In the peripheral organs the pressure drop in the capillaries is relatively minor. In the lung, one-third or more of the total pressure drop from pulmonary valve to left atrium may reside in the capillaries. Under negative venous pressure (below atmospheric) a "waterfall phenomenon" (flow limitation) occurs in the pulmonary capillaries, but not in peripheral capillaries.

Blood flow in the heart and lung is usually referred to as *central circulation*; that in other organs is said to be *peripheral*, which includes coronary and bronchial circulations. The systolic arterial blood pressure in peripheral organs is higher than that in the lung. The volume of the tissues served by the systemic capillaries is much larger than that surrounding the pulmonary capillaries. The systemic capillaries are embedded in tissues, whereas the pulmonary capillaries are balanced against the alveolar gas. These are additional differences of pulmonary and systemic circulations.

But circulation is circulation. Most of the features of peripheral and pulmonary circulations are the same. In both circulations it is often convenient to consider features at different levels of sizes. For example, one may wish to discuss blood flow in the whole lung, or that in a lobule, or an acinus, or a single capillary. One may wish to analyze the cellular events in terms of cell membrane, cytoskeleton, actin molecules, and so on. To express the relationship between one level of size to another, the terms *macro* and *micro* are convenient. The theme of this chapter is to illustrate some of the connections between micro- and macromechanics of the lung.

Basic Morphometric and Rheological Data

The mammalian respiratory system consists of three trees: the airway tree, the arterial tree, and the venous tree (Fig. 13.1). The airway tree consists of the nose, mouth, larynx, trachea, bronchi, bronchioles, respiratory bronchioles, alveolar ducts, and alveoli. The walls of the alveoli are capillary blood vessels. The arterial tree consists of pulmonary arteries, arterioles, and capillaries. The venous tree consists of capillaries, venules, veins, and left atrium. Morphological data of these trees of man and animals have been collected through the ages by many people [see summaries and discussions by Horsfield (1978), Horsfield and Gordon (1981), Krahl (1964), Miller (1947), Weibel (1963), Yen et al. (1983, 1984b), and others].

To understand pulmonary circulation, one needs not only the morphological data, but also the constitutive equations that describe the stress-strain-history relationship of the materials of the system. Even in a simplified theory in which blood vessels are treated as tubes, we need data on pressure-diameter relationship. Data on the distensibility of pulmonary arteries, veins, and capillaries have been collected by Caro et al. (1967), Patel et al. (1963), Sobin et al. (1972, 1978), and Yen et al. (1980, 1981) using a variety of methods: isolated vessel segments, x-ray angiography, and catalyzed polymer perfusion followed by hardening at known fluid pressure, tissue corrosion, and stereological morphometry.

Due to the effort of Sobin, Yen, and their colleagues, a full set of morphometric and distensibility data on all generations of pulmonary vessels of the cat is now available. These data are summarized in Fung (1984).

One needs also the rheological data of the blood. For the analysis of pulmonary blood flow, it is sufficient to know the *apparent coefficient of viscosity* of the blood. A series of studies was done by Lee (1969), Lee and Fung (1968), and Yen and Fung (1973) to clarify the flow of plasma and red blood cells in pulmonary alveolar capillary sheets. The results show that a linear relationship between the mean velocity of flow and pressure gradient holds. The effects of sheet geometry, the cell diameter/sheet thickness ratio, and the cell versus sheet wall deformability ratio have been identified.

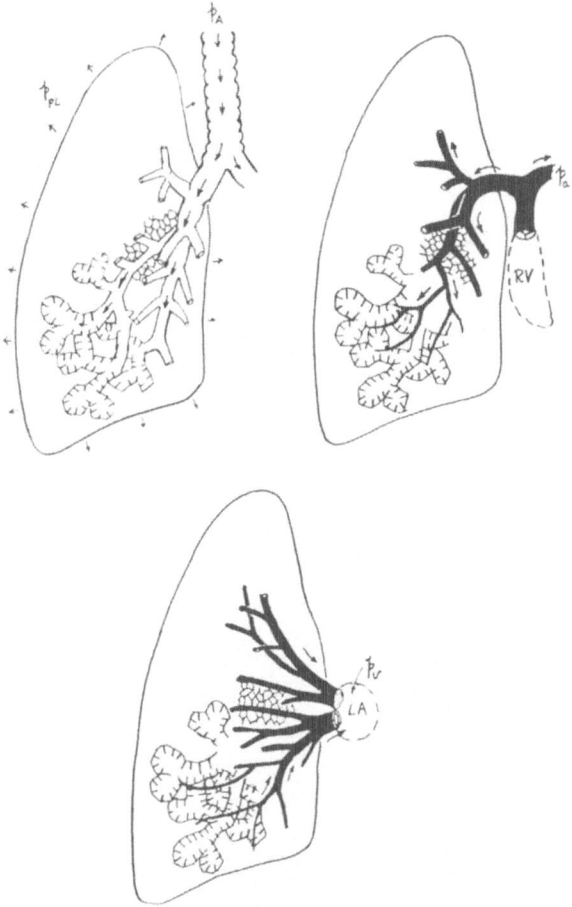

FIGURE 13.1. A schematic drawing of the bronchial, arterial, and venous trees in the lung.

These data provide a basis on which a theory of pulmonary blood flow can be built and validated. Among other things, three pieces of information especially relevant to the discussions below are shown in Figures 13.2 through 13.4. Figure 13.2 shows the distensibility of the pulmonary capillary sheet in the direction perpendicular to the sheet. This dimension, called *thickness* and denoted by h, is defined as the mean distance between the endothelial surfaces averaged over an alveolar wall. The value of h varies with the transmural pressure, ΔP, which is the difference of blood pressure and alveolar gas pressure. When ΔP is positive and less than an upper limit ΔP_u, h is a linear function of ΔP:

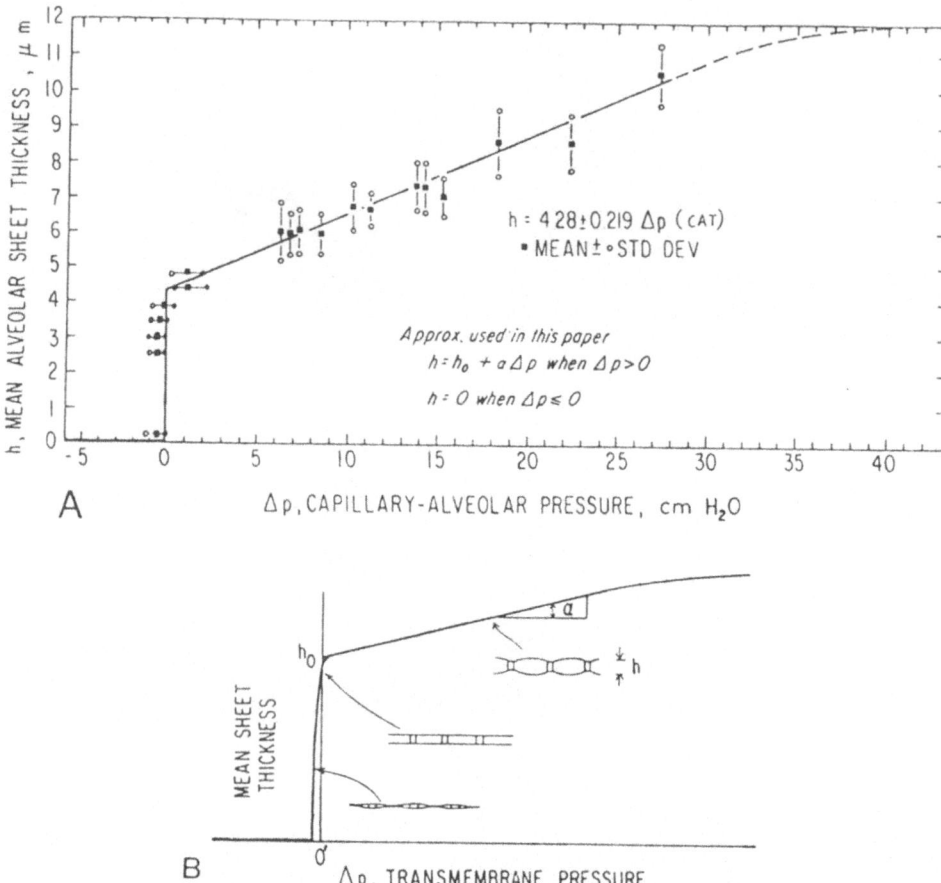

FIGURE 13.2. *A*: Sheet thickness-pressure relationship of the cat. Data can be approximated by a discontinuous curve that is composed of four line segments: a horizontal line $h = 0$ for ΔP negative, which jumps to $h = h_0$ at ΔP slightly greater than 0, then continues as a straight line for positive ΔP until some upper limit is reached, beyond which it bends down and tends to a constant thickness. *B*: The elastic deformation of the alveolar sheet is sketched for three conditions: $\Delta P < 0$, $\Delta P = 0$, and $\Delta P > 0$. The relaxed thickness h_0 of the sheet is equal to the relaxes length of the posts. Under a positive internal (transmural) pressure, the thickness of the sheet increases; the posts are lengthened; the membranes deflect from the planes connecting the ends of the posts; and the mean thickness becomes h. (From Fung and Sobin, 1972a); by permission of the American Heart Association, Inc.)

$$h = h_0 + \alpha \Delta P \tag{13.1}$$

For $\Delta P > \Delta P_u$, h increases at a slower rate with ΔP, and tends asymptotically to a constant when $\Delta P \gg \Delta P_u$. h tends to h_0 when $\Delta P \to 0$ from the positive side. When ΔP becomes negative, h drops down to zero rapidly. h remains zero when ΔP is smaller than about 1 cm H_2O. This tendency to collapse at negative transmural pressure is the basic cause of the "waterfall phenomenon" to be discussed later.

What is the distensibility of the pulmonary capillaries in the plane of the alveolar wall? The answer is that it is very small. With respect to changes of blood pressure, the pulmonary capillaries appear to be rigid in the plane of the alveolar wall. The alveolar wall can be distended by the transpulmonary pressure, $P_A - P_{PL}$, (the alveolar gas pressure minus the intrapleural pressure); but not by the blood pressure.

The directional difference of the distensibility of the pulmonary capillaries is one of the reasons to identify the network of pulmonary capillaries as a *sheet*. A sheet is characterized by its area and thickness, unlike a cylindrical tube, which is characterized by its diameter.

Figure 13.3 shows a plot of the diameter of pulmonary veins as a function of $P - P_A$ (blood pressure minus alveolar gas pressure). Note that these curves are straight lines, unlike the familiar exponential curves of the aorta, vena cava, and peripheral arterioles and venules. Note especially that they go through the point $P - P_A = 0$ without changing slope. In other words, the pulmonary veins would not collapse at negative transmural pressure, at least when $\Delta P > -23$ cm H_2O, the range tested by these experiments. If the abscissa is changed to $P - P_{PL}$, a similar conclusion is obtained. In contrast, if this were a vena cava or a small vein of a peripheral organ, the curve would have turned downward very rapidly when $P < P_A$. Peripheral veins subjected to a negative transmural pressure would buckle and collapse, decrease its cross-sectional area rapidly for a slight increase of compression. Not so for pulmonary veins.

The explanation of this stability of pulmonary veins and arteries lies in the support the pulmonary vessels receive from the attached interalveolar septa. The pulmonary veins and arteries are embedded in an elastic medium.

Figure 13.4 is a photograph obtained by Sobin et al. (1983) of a pulmonary arteriole of the rat in normal condition. Figure 13.5 is a photograph of a similar arteriole after the rat was exposed to hypoxia at PO_2 of 40 mm Hg for 24 h. Note the swelling of the endothelium, the formation of blebs, and the edema of subendothelial space. The rapid morphological change is impressive. Sobin et al. (1983) followed the changes in rat lung from 15 min to 1 yr of exposure to hypoxia. These photographs suggest that one of the reasons for hypertension in hypoxia is the swelling of the intima of the arterioles. They show how rapidly remodeling of the blood vessel takes place following the onset of hypoxia.

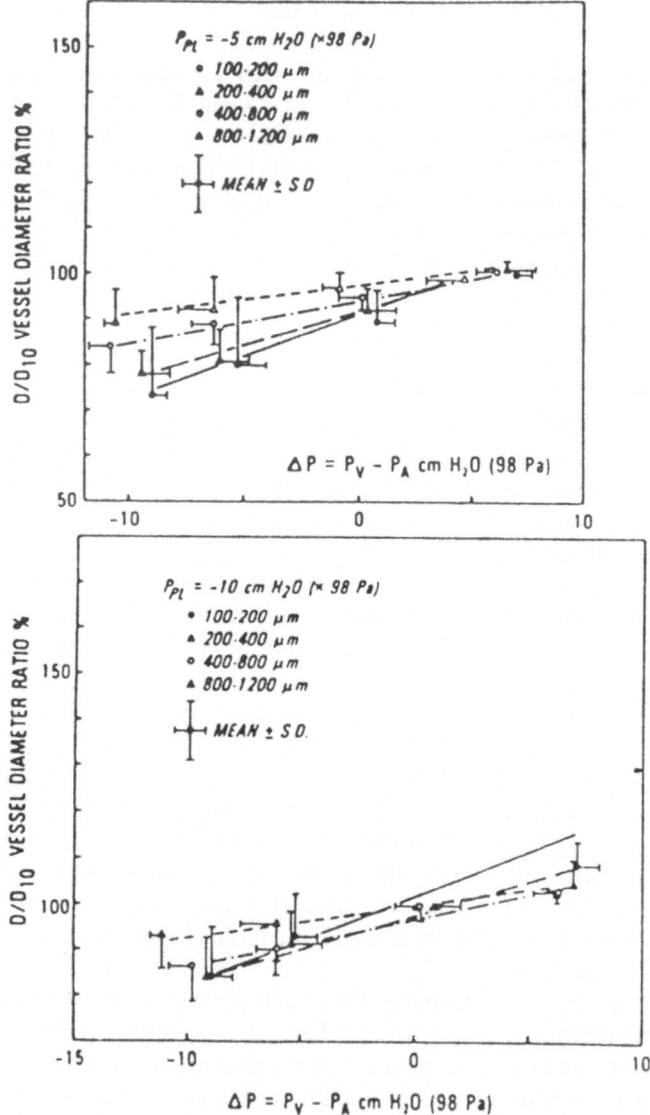

FIGURE 13.3. Percentage change in diameter of pulmonary veins of the cat as a function of the blood pressure. The airway pressure P_A, is zero. The values of the pleural pressure, P_{PL}, are noted in the figure. The vessel diameter is normalized against its value when $P_v - P_A$ is 10 cm H_2O, at which the vessel cross-section is circular. The nominal size of a vessel is its diameter at $P_v - P_A = 10$ cm H_2O. (Reprinted with permission from Transactions of the Asme, *J Biomech Eng* Yen and Foppiano, 1981.)

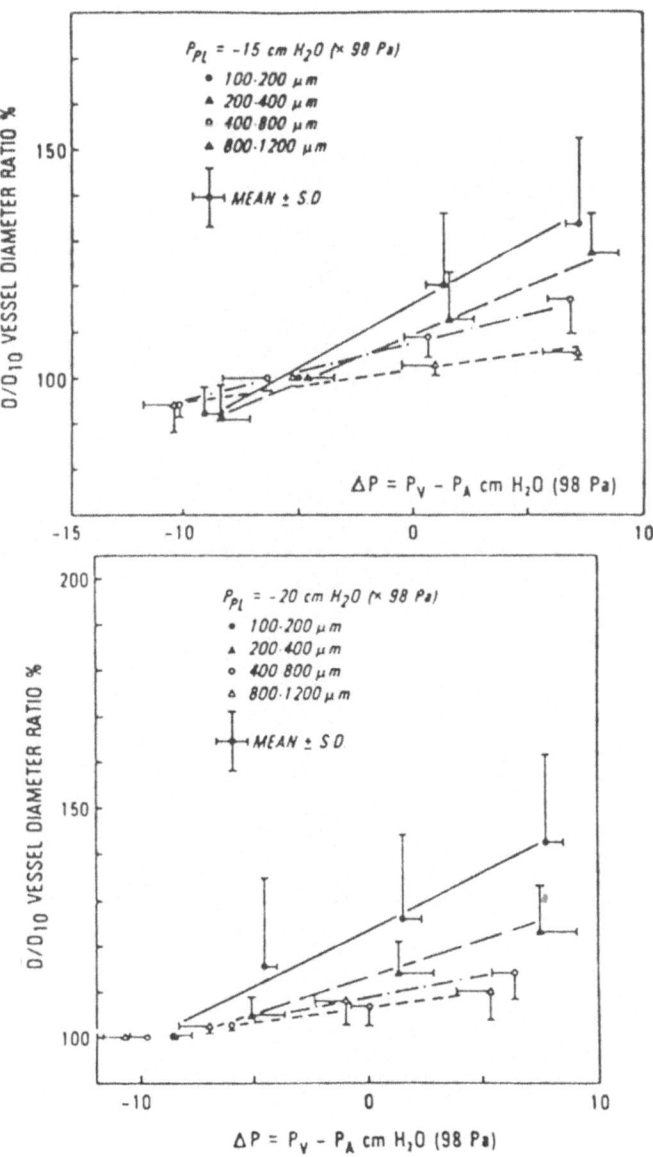

FIGURE 13.3 (*Continued*)

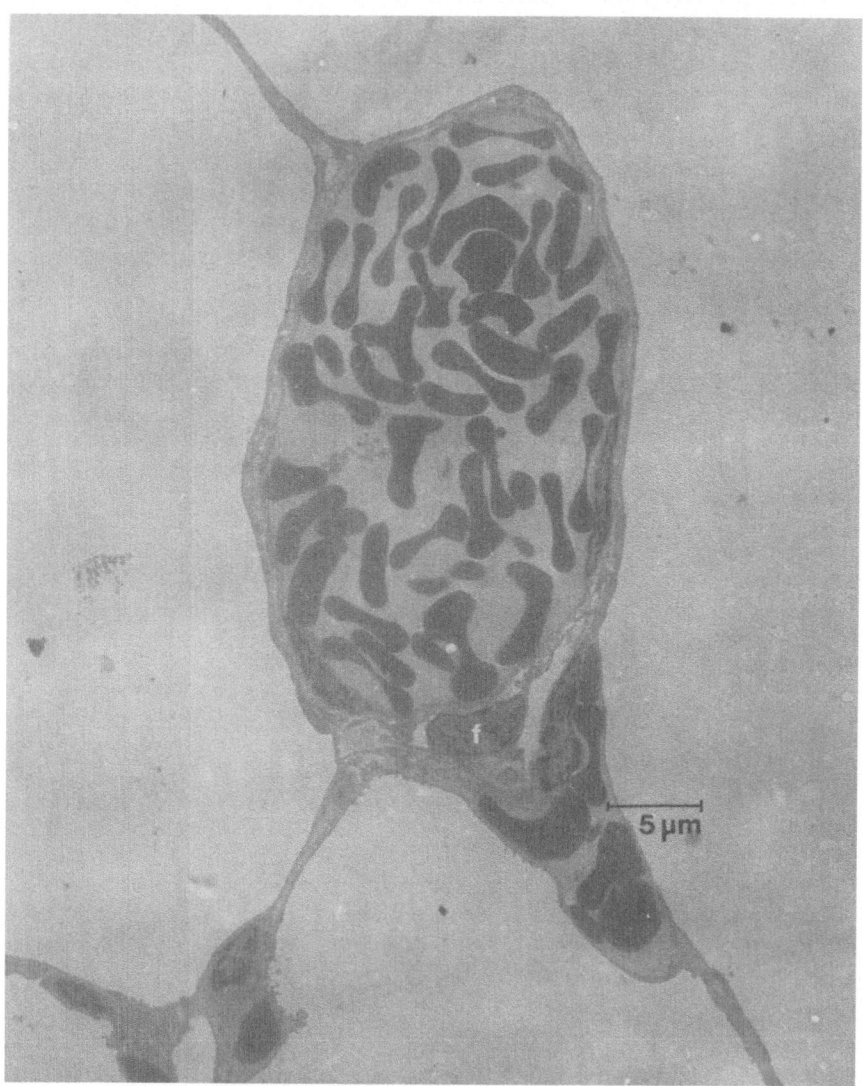

FIGURE 13.4. Electron micrograph of a small arteriole in normal rat lung. Note relatively thin arteriolar wall and a fibroblast (f). (From Sobin et al. 1983.)

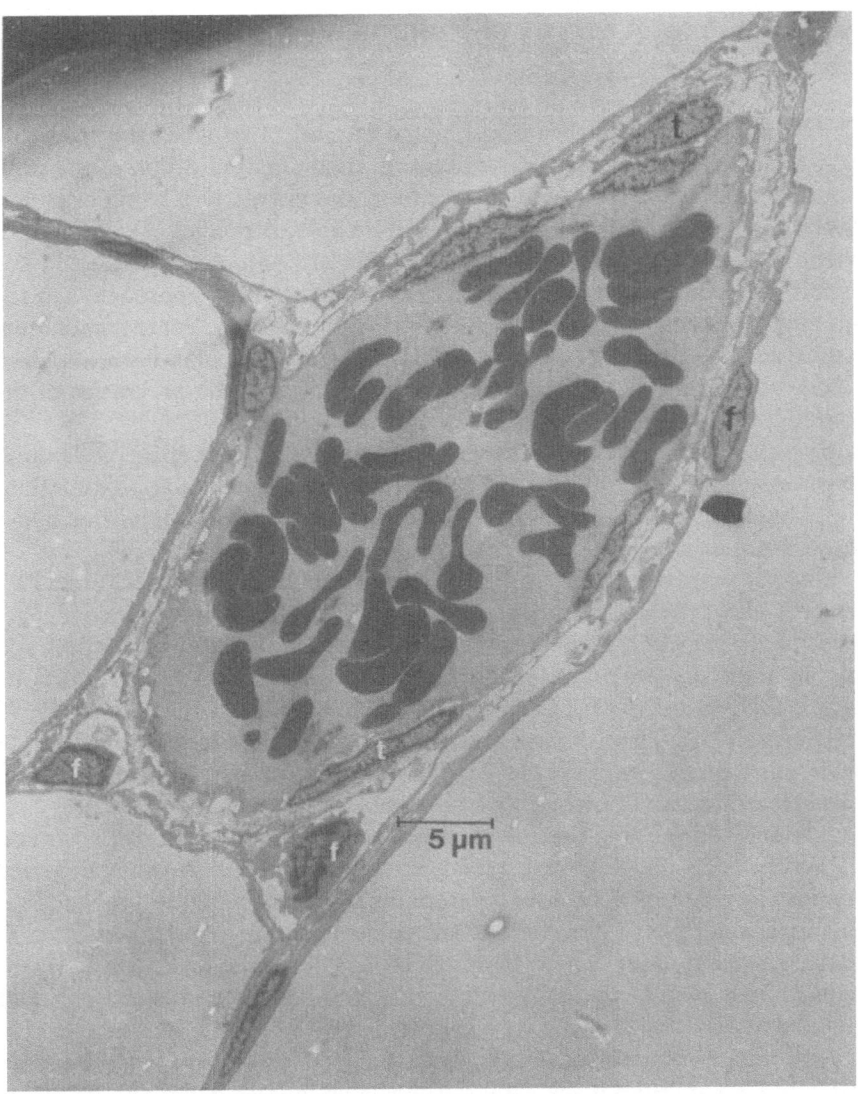

FIGURE 13.5. Electron micrograph of an arteriole after 24 h of 10% O_2. Several fibroblasts (f) can be seen in interstitial areas. t, Transitional cell. (From Sobin et al., 1983.)

Mathematical Analysis of Pulmonary Alveolar Blood Flow

The basic hypothesis of theoretical biomechanics is that when the geometric details and material properties are known, then the application of the principles of conservation of mass, momentum, and energy to a living organism under specified boundary conditions will be able to predict the mechanical phenomena occurring in the organism accurately, without any other ad hoc hypotheses. The emphasis is laid on the simplicity of the approach, and the minimal number of hypotheses. Other factors, such as the nerve stimulation, muscle contraction, oxygen consumption, CO_2 release, metabolism, and drug effects, are considered to influence the geometry and the properties of the materials involved.

Hypotheses are tested by experimental validation of theoretical predictions. Experimentation and theoretical calculations may involve approximations and uncertainties that are ad hoc. Evaluation of the effects of ad hoc hypotheses is a necessary part of biomechanics.

Under this general principle we approach pulmonary circulation. Pulmonary blood flow shares with the peripheral circulation the same phenomena of ventricular pumping, windkessel, wave propagation, reflection, and refraction. Wave propagation in pulmonary arteries has been studied by Bergel and Milnor (1965), Caro (1967), Milnor (1972), Pollack et al. (1968), and Wiener et al. (1966). Waves are significant in large arteries and veines. In pulmonary capillaries pulsatile flow is recognizable but can be treated as quasi-static. In the following, we shall consider the time-averaged pressure-flow relationship. Using the energy-balance equation (Fung, 1984, pp. 9 and 15; Pedley et al., 1977, p. 181), averaging it over a period of oscillation, one obtains an equation that relates the static and dynamic pressure, gravitational potential, and energy dissipation. The same equation is obtained by considering a steady flow. Hence, although steady flow does not occur in mammalian lung in life, it is still meaningful to consider it theoretically and experimentally.

With this understanding, Fung and Sobin (1972a, b) formulated a theory of blood flow in pulmonary capillary sheets as follows. Since the Reynolds number of sheet flow is very small ($\sim 10^{-3}$ to 10^{-2}), inertial forces can be ignored. The equation of motion combined with blood rheology leads to the equations

$$\frac{\partial P}{\partial x} = -\frac{\mu k f}{h^2} U, \qquad \frac{\partial P}{\partial y} = -\frac{\mu k f}{h^2} V \qquad (13.2)$$

where P is the blood pressure, x, and y are rectangular coordinates in a capillary sheet, U and V are the corresponding local average velocities in the sheet, h is the local sheet thickness, μ is the apparent coefficient of viscosity of the blood, and k and f are two dimensionless parameters (k is a function of the ratio of sheet width to sheet thickness (in practice $k \simeq 12$), and f is a function of the hematocrit, cell rigidity/wall rigidity ratio, post diameter/h, interpostal distance/h, and postal pattern (in practice $f = 3$ to 5 depending on the sheet geometry; see Yen and Fung, 1973). The sheet elasticity is embodied by Eq. 13.1:

$$h = h_0 + \alpha(P - P_A)$$

The conservation of mass is expressed by the equation

$$\frac{\partial(hU)}{\partial x} + \frac{\partial(hV)}{\partial y} = 0 \tag{13.3}$$

if edema and pulsation are ignored. Eliminating P, U, and V from Eqs. 13.1, 13.2, and 13.3 and assuming μ, k, and f to be constants, we obtain

$$\left(\frac{\partial^2}{\partial x^2} + \frac{\partial^2}{\partial y^2}\right)h^4 = 0 \tag{13.4}$$

Solving this equation—putting in the boundary conditions that at the artriole supplying an alveolar sheet the pressure is P_{art} and the thickness is h_{art}, whereas at the venule draining the flow in the sheet the pressure and thickness are P_{ven} and h_{ven}, respectively—we obtain, on integrating over all streamtubes in the sheet, the total flow (Fung, 1984, p. 331):

$$\text{Flow} = \frac{1}{C}(h_{art}{}^4 - h_{ven}{}^4) \tag{13.5}$$

where

$$C = \frac{4\mu k f \bar{L}^2 \alpha}{SA} \tag{13.6a}$$

$$h_{art} = h_0 + \alpha(P_{art} - P_A) \tag{13.6b}$$

$$h_{ven} = h_0 + \alpha(P_{ven} - P_A) \tag{13.6c}$$

where $\bar{L}$ is the average length of the streamtubes in the sheet, A is the sheet area, S is the "vascular space/sheet volume" ratio (about 0.89 to 0.91 for man, dog, and cat) and α is, again, the compliance constant of the alveolar sheet (Eq. 13.1). Eq. 13.5 gives the full solution in the region of linear P vs h relationship:

$$0 \leq P_{ven} - P_A \leq P_{art} - P_A \leq P_U \tag{13.7}$$

that is, in the so-called *zone 3* condition. The last inequality sign limits the applicability of the formula to a region of modest blood pressure in zone 3. If the blood pressure is so high that

$$P_{ven} - P_A > P_U \qquad and \qquad P_{art} - P_A > P_U \tag{13.8}$$

then the compliance α decreases, as shown in Figure 13.2. When the compliance is very small, the difference of the thicknesses at the arteriole and venule is small. Then Eq. 13.5 becomes

$$\text{Flow} = \frac{1}{C}(h_{art} - h_{ven})(h_{art}^3 + h_{art}^2 h_{ven} + h_{art} h_{ven} + h_{ven}^3)$$

$$\doteq \frac{1}{C}\alpha(P_{art} - P_{ven})4h_{art}^3 \tag{13.9}$$

In this case, the compliance constant disappears from the product α/C. If

$$P_{ven} < P_{art} < P_A \tag{13.10}$$

then the condition is said to be *zone 1*. In zone 1 the capillaries are collapsed and the flow is zero. In a standing man, the apex of the lung is in zone 1.

Waterfall Phenomenon in Zone 2

The condition

$$P_{ven} < P_A < P_{art} \tag{13.11}$$

defines the zone 2 condition. In a standing man, a region below zone 1 and above zone 3 (sufficiently high above the left atrium) is zone 2. In this zone, h_{ven} could be much smaller than h_{art}; then the last term in Eq. 13.5 is negligible and

$$\text{Flow} \doteq \frac{1}{C}h_{art}^4$$

$$\doteq \frac{1}{C}[h_0 + \alpha(P_{art} - P_A)]^4 \tag{13.12}$$

The flow is then independent of the downstream condition, in analogy with a waterfall.

The waterfall phenomenon in lung was discovered by Dr. Permutt (Permutt et al., 1962, 1963, 1969). In earlier literature there was a debate as to where on the vascular tree the "waterfalls" or "sluicing gates" are located. At that time the prevailing concepts were that capillaries are rigid, veins are collapsible, and arterioles are muscular and vasoactive. Hence the sluicing gates were searched for in veins and arterioles. Fung and Sobin (1972b), however, showed that pulmonary capillaries are collapsible, whereas Fung et al. (1983) showed that pulmonary arteries and veins (and venules) will not collapse under negative $P_{ven} - P_A$. Hence the sluicing gates must be located at the venular ends of the capillaries, and Eq. 13.12 holds.

Fung (1984) pointed out that one of the solutions of Eq. 13.4 is

$$h = 0 \tag{13.13}$$

which can occur if an area of the alveolar sheet is collapsed. Furthermore, the multifaceted alveolar structure of the lung is such that areas of collapsed

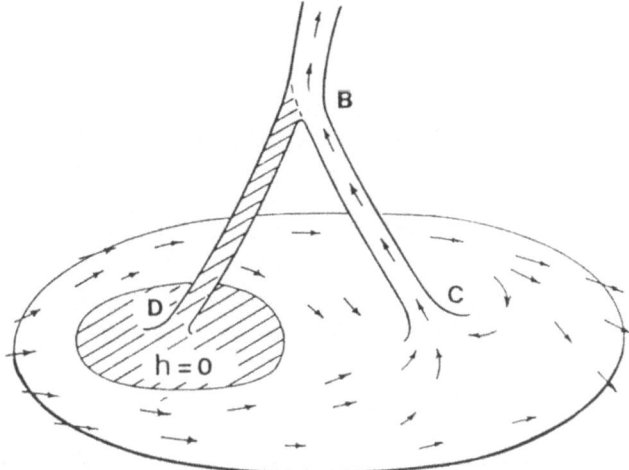

FIGURE 13.6. A schematic drawing showing a possible condition of flow from an alveolar sheet into a venous tree. Two terminal venules (of order 1) intersect the sheet. The blood pressure at point B is smaller than the alveolar gas pressure (i.e., $P - P_A < 0$). Blood flows from the sheet into the venule BC; the junction C is thus a sluicing gate. A portion of the sheet around the terminal D of the venule BD is, however, closed. The vascular sheet thickness is zero in the shaded region, where there is no flow. The blood pressure at D is equal to that at B, whereas that at C is greater than that at B. (From Fung, 1984.)

alveolar sheets can be embedded in other alveolar sheets that are open and perfused. The concept is shown in Figure 13.6. Here the collapsed area is shown blackened. The boundary conditions for the solution in Eq. 13.13 are satisfied. In the opened area there is flow, and Eq. 13.5 applies. The total perfused area S in the constant C of Eq. 13.6 is equal to the total lung alveolar sheet area minus the part that is collapsed ($h = 0$). Hence to determine the flow in zone 2 condition, we must estimate the alveolar sheet area that is closed.

Open and Closed Alveolar Capillary Sheets in Zone 2

The estimation of the area of the open capillary sheets is helped by several pieces of new information.

First, Fung (1984) and Fung and Zhuang (1986) investigated the details of flow through the sluicing gate, taking into account the effect of tension in the alveolar wall and the local curvature of the membrane, as well as the Stokes flow equation. It is shown that as P_{ven} is decreased below P_A, the flow in the gate is increased while the gate narrows further. This theory replaces the "cusp" argument put forward in Fung and Sobin (1972a).

Second, Fung and Yen (1986) investigated the stability of a partially collapsed interalveolar septum, illustrated in Figure 13.7B, by the method of potential energy. This sheet in Figure 13.7B is open at the left end, closed on the right end, and partially closed in part of the sheet. Let the strain energy of the bent wall and the compressed posts be denoted by W_D, the work done by the alveolar gas and blood in creating a partially collapsed sheet be denoted by W_P, whereas the free energy of the surface be denoted by W_{CB}, (the subscript CB stands for "chemical bond"). Then if a small increment of the area of contact (adhesion), δA_c, occurs, small changes in W_{CB}, W_D, W_P would occur. If the sum $\delta W_{CB} + \delta W_D + \delta W_P$ were negative, there would be a tendency for the contact area to increase. At equilibrium, we have

$$\frac{\partial}{\partial A_c}(W_{CB} + W_D + W_P) = 0 \qquad (13.14)$$

by which the area of contact A_c can be determined. Furthermore, the collapsed area is *stable* if

$$\frac{\partial^2}{\partial A_c{}^2}(W_{CB} + W_D + W_P) > 0 \qquad (13.15)$$

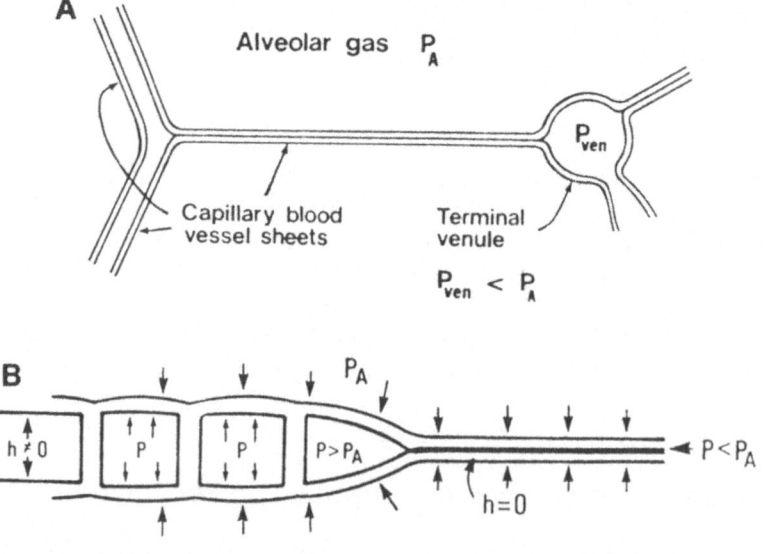

FIGURE 13.7. *A*: Sheet of pulmonary capillary blood vessels (an interalveolar septum) is collapsed. Collapse is arrested by two open septa shown on left. *B*: Septum with one part adjacent to a venule is collapsed. Deformation of walls is illustrated. Collapsed lumen is represented by a thicker line to remind reader that membranes are separated by "posts," which do not collapse, and thus some very small areas around posts must remain open. This type of collapse is unstable, and a collapse, once started, will continue to spread until arrested by something else. P_A, alveolar pressure; P_{ven}, venule pressure; h, capillary sheet thickness. (From Fung and Yen, 1986.)

It is *unstable* if

$$\frac{\partial^2}{\partial A_c^2}(W_{CB} + W_D + W_P) < 0 \tag{13.16}$$

When the details were worked out, Fung and Yen (1986) found that Eq. 13.16 prevails. Hence if an interalveolar septum in zone 2 starts to collapse, it will continue to collapse until the whole septum is collapsed. The result is illustrated in Figure 13.7A. In this figure, the left end of the collapsed septum is adjoined to two open septa. If one of the open septam started to collapse, it would continue to do so until the whole septum is collapsed, and so on.

This theoretical result is corroborated by histological evidence obtained earlier by Warrell et al. (1972), reproduced in Figure 13.8. Dog lung at zone 2 flow condition was quick frozen by pouring liquid Freon cooled to liquid nitrogen temperature. Specimens were taken within several millimeters below the pleura, freeze dried, and processed to preserve the geometry. In Figure 13.8, it is seen that several interalveolar septa are open and several are collapsed, trapping a few red blood cells in them. Note that those septa that are collapsed did so as a whole.

Summarizing, we found that in zone 2 condition, a number of interalveolar septa (sheets) connected to the venules may be collapsed, while the remaining sheets are open. Sluicing occurs in the open sheets, in which the gates are located at the junctions with the venules.

A given interalveolar septum may be open or may be closed. To reopen a closed sheet requires additional energy to recreate the free surface. To make

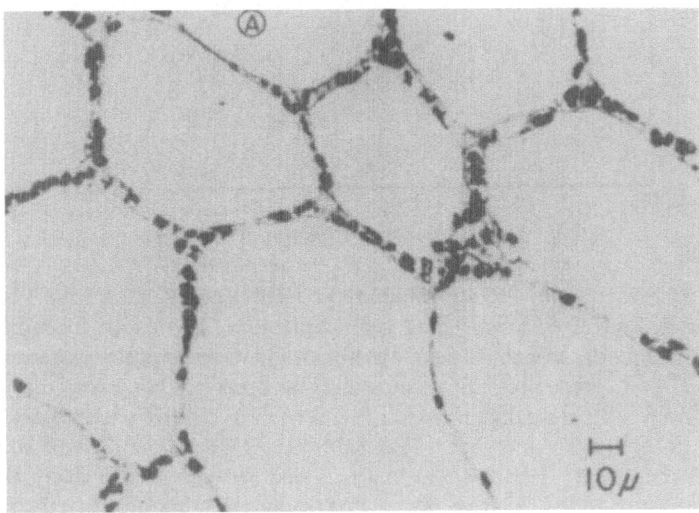

FIGURE 13.8. Histological micrograph of dog lung in zone 2 condition. Specimen obtained by quick freezing, then fixed and dried at critical temperature. (From Warrell et al., 1972.)

sure that every sheet is open in a lung, one should impose a large flow while the lung is in zone 3 condition. If the venous pressure is continuously reduced from zone 3 condition to zone 2 condition, more sheets will be collapsed as the venous pressure is lowered. Since the collapsing is stochastic, Fung and Yen (1986) assumed a normal probability

$$\frac{A_c}{A} = F(1 - e^{-\Delta P^2/2\sigma^2}) \tag{13.17}$$

to relate the collapsed area, A_c, with the pressure $\Delta P = P_{ven} - P_A$ when $\Delta P < 0$. Here A is the total anatomical alveolar sheet area and F and σ are constants. This formula was validated by experimental results such as that illustrated below in Figure 13.9. The physical meaning of F is the largest fraction

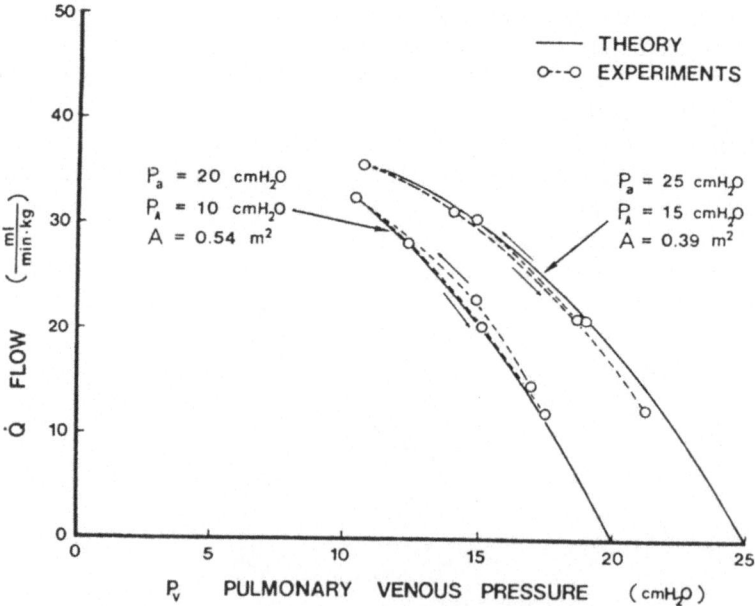

FIGURE 13.9. Pressure-flow relationship in two right lungs of cat in zone 3 condition. Each lung was tested first by cycling venous pressure (P_v) to zone 2 condition. The value of P_v at which flow reached peak was noted, and was interpreted as corresponding to the condition venule pressure (P_{ven}) = alveolar pressure (P_A). Same lung was then tested by cycling P_v in a range in which $P_{ven} > P_A$. For example, in the case in which $P_A = 15$ cm H_2O, it was found that $P_v = 8.0$ cm H_2O when flow ($\dot{Q}$) reached the peak value of 38 ml/min/kg. Hence, when P_v was cycled between 22 and 10 cm H_2O, flow condition was in zone 3; that is, $P_{ven} > P_A$ throughout, although $P_v < P_A$ in part of cycle. Lower curve was from another experiment, which had a $\dot{Q}_{max}$ of 44 ml/min/kg when P_v was 2.4 cm H_2O. Theoretical curves were drawn with alveolar wall surface area A (half lung) indicated in figure. Good fit is obtained by adjusting the values of A. P_a, arterial pressure. (From Fung and Yen, 1986.)

of the alveolar sheet that will be collapsed when the pulmonary venous pressure is lowered indefinitely. This number is theoretically predictable when the relationship between the arteries, veins, alveolar ducts, and alveoli are known. This requires a model of the alveolar ducts. We shall discuss our model in "Validation of the Flow Model," below. In Fung and Yen (1986), a theoretical derivation of F based on a pentagonal dodecahedral model of the lung parenchyma yielded F = 0.156. Identification of prediction with experimental data yields F = 0.104 $\pm$ 0.016 (SE), σ = 4.45 $\pm$ 0.45 (SE).

Following Eq. 13.17, and the fact that in zone 2 blood flow exists in those capillary sheets that are *open*, we see that Eqs. 13.5 and 13.6 remain valid if the sheet area A in those equations is replaced by the *open* area $A - A_c$. It follows that in zone 2

$$\text{Flow} = \left(1 - \frac{A_c}{A}\right)\frac{SA}{4\mu k f L^2 \alpha}(h_{art}^4 - h_{ven}^4) \tag{13.18}$$

with A_c/A given in Eq. 13.17.

Synthesis of Micro- and Macrocirculation in the Lung

The circulation of the whole lung is the sum of flow through all the segments. But an analysis of a circuit can be done only if one knows the details of the circuit. Anatomical data in the form of Strahler system (see "Basic Morphometric and Rheological Data," above) do not specify the circuit completely. It is, however, consistent with anatomical observation and these data to assume that the vascular tree can be treated with the successive generations in series, but within each generation several possible topological arrangements can be specified. Using this simplifying assumption, Zhuang et al. (1983) presented a detailed analysis of flow in zone 3 condition; Fung and Yen (1986) presented a detailed analysis of flow in zone 2 condition. Experimental validation of these results were published in Yen et al., (1983, 1984a, 1984b, 1986).

Whereas theoretical prediction and experimental validation are discussed in the next section, the theoretical procedure of synthesis will be outlined presently. Begin with the capillaries, in which the pressure-flow relationship is given by Eqs. 13.5 through 13.18. Consider the first generation of the arterioles and venules, in which the Reynolds and Womersley numbers are smaller than unity. In these vessels the geometry is tubular, the flow in each segment is constant, the average velocity varies inversely with the cross-sectional area, and the pressure drop follows Poiseuille's formula:

$$\frac{dP}{dx} = -\frac{8\mu}{\pi a^4}\dot{Q} \tag{13.19}$$

where $\dot{Q}$ is the volume-flow rate, P is the pressure, x is the axial length, μ is the apparent coefficient of viscosity, a is the radius. The radius-pressure relationship presented in Figure 13.2 can be represented by the equation

$$a = a_0 + \frac{\alpha}{2}P \tag{13.20}$$

in which α is the *compliance constant* and a_0 is the radius when the transmural pressure is zero. Combining Eqs. 13.18 and 13.19, integrating with respect to x, and using the end conditions that $a = a(0)$ when $x = 0$, and $a = a(L)$ when $x = L$, we obtain

$$\dot{Q} = \frac{\pi}{20\mu\alpha L}\{[a(0)]^5 - [a(L)]^5\} \tag{13.21}$$

which gives the pressure-flow relationship in these small vessels.

In larger vessels with Reynolds number greater than 1, the effect of kinetic energy change on the pressure drop must be considered. As explained in the Introduction, we shall only consider steady flow. In a tubular vessel with one-dimensional steady flow, the equation of motion is

$$\rho q \frac{\partial q}{\partial x} = -\frac{dP}{dx} - \frac{8\mu}{\pi a^4}\dot{Q} \tag{13.22}$$

Here ρ is the density of the blood, q is the velocity in the tube averaged over the cross-section, and μ is the apparent coefficient of viscosity, which are corrected for turbulence, if any. The equation of continuity is

$$\pi a^2 q = \dot{Q}, \tag{13.23}$$

a constant. The pressure-radius relationship is Eq. 13.20. Combining Eqs. 13.20, 13.22, and 13.23, eliminating P, and integrating, we obtain

$$a^5 - \frac{5\rho\alpha\dot{Q}^2}{\pi^2}\ln a = -\frac{20\mu\alpha}{\pi}\dot{Q}x + \text{const.} \tag{13.24}$$

Applying the end conditions $a = a(0)$ at $x = 0$ and $a = a(L)$ at $x = L$, we obtain

$$\dot{Q} - \left[\frac{\rho}{4\mu\pi L}\ln\frac{a(L)}{a(0)}\right]\dot{Q}^2 = \frac{\pi}{20\mu\alpha L}\{[a(0)]^5 - [a(L)]^5\} \tag{13.25}$$

The second term is a correction of Eq. 13.21 for the effects of the inertial force. This correction is in general quite negligible in pulmonary vessels.

Now we can synthesize all the segments into a circuit. Flow in each segment is given by Eq. 13.5 or 13.9 for capillaries in zone 3, Eq. 13.18 for capillaries in zone 2, Eq. 13.21 for small arteries and veins, and Eq. 13.25 for large arteries and veins. At the junctions of the vessels of successive generations a finite jump of velocity occurs because of a step change in total cross-sectional area. The average velocity q_n in a pulmonary artery of order n is greater than that in the next order q_{n-1} (the order number being counted from capillary upward). Hence, according to Bernoulli's equation, there is a sudden *jump* of pressure equal to

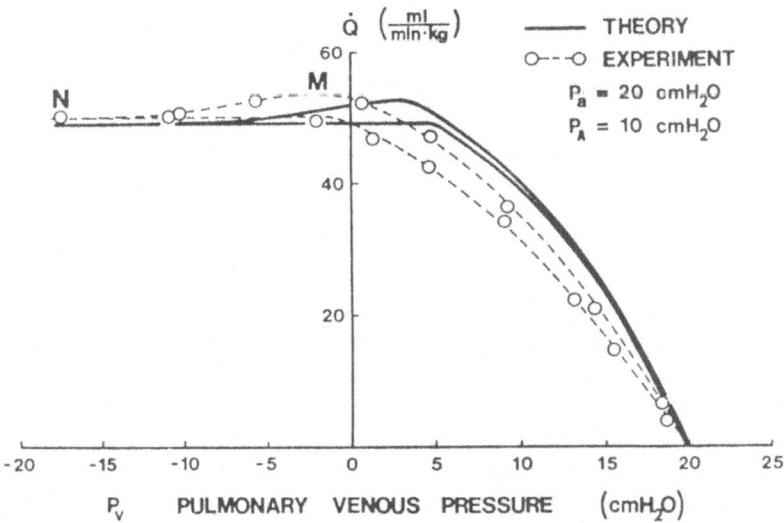

FIGURE 13.10. Comparison of theoretical and experimental results in pulmonary blood flow. Theoretical curves were computed with morphometric and elasticity data given in Fung (1984) for right lung of cat, and $A = 0.84$ m^2 for half lung. In zone 2 condition, Eq. 13.17 was used to compute collapsed alveolar surface area (A_c). Curve fitting yields $\sigma = 5.3$ and F $= 0.093$. In return stroke, alveolar surface area is taken to be $A - A_c$; that is, collapsed area is deducted from initial value of total alveolar wall surface area of right lung under the assumption that collapsed alveolar sheets are not reopened until pulmonary arterial pressure is increased in zone 3 condition. Value of A was so selected that a reasonable agreement between theory and experiment is obtained. (From Fung and Yen, 1986.)

$$\tfrac{1}{2}\rho q_n^2 - \tfrac{1}{2}\rho q_{n-1}^2 \tag{13.26}$$

Similarly, at a junction of pulmonary veins of orders $n-1$ and n, there is a sudden *drop* of pressure equal to

$$\tfrac{1}{2}\rho q_n^2 - \tfrac{1}{2}\rho q_{n-1}^2$$

In addition, there is the effect of entry flow and exit flow, which, however, can be taken into account by correcting the apparent coefficient of viscosity of the blood.

Calculations using detailed anatomic and rheological data following the method just outlined were done by Zhuang et al. (1983). Some results in the cat lung are illustrated in Figures 13.9 and 13.10.

Validation of the Flow Model

Yen et al. (1985, 1986) and Fung and Yen (1986) perfused cat lung and obtained the experimental results shown in Figures 13.9 and 13.10. Figure 13.9 refers to zone 3 condition. Figure 13.10 refers to zone 2 condition. In these experiments the pressures in the airway and the largest artery were fixed, whereas the pressure in the left atrium was varied, first from a higher value downward, reaching a minimum, then back upward. The lungs were "preconditioned" by giving it a few cycles of large flow in zone 3 condition to open up all vessels. Note that in zone 2 condition (Fig. 13.10), the flow reached a peak at a certain value of P_{ven}, then decreased with further decrease of P_{ven}. A hysteresis loop exists under cyclic change of P_{ven}.

These experimental results are compared with theoretical predictions in Figures 13.9 and 13.10. In the theoretical calculation, it is necessary to know the total area of the alveolar sheets. This area was measured from histological sections of a given lung by stereological methods, but the standard deviations of the measured values were large. The origin of the large standard deviation is partly due to biological variations from one animal to another and from one location in the lung to another, and partly due to finite sampling, and distortion caused by histological preparations, especially shrinking by fixation agents. Since the available data for the total area (A in Eqs. 13.6 and 13.18) are not very precise, we have selected a value of A that makes the theoretical prediction coincide with an experiment result at one point on the P-$\dot{Q}$ curve as noted in Figures 13.9 and 13.10. This value of A was found to lie in the range of experimentally determined values. If A is varied, the theoretical curve would move up and down since the flow $\dot{Q}$ is directly proportional to A. Thus it is seen that the experimental trend is well predicted by the mathematical model.

Relationship of Pulmonary Arterioles, Venules, Alveolar Ducts, and Alveoli

To gain a deeper understanding of the lung one needs to know the coupling between circulation and respiration at the microscopic level. For example, to evaluate the constant F in Eq. 13.17 one must know the geometric relationship between arterioles, venules, and interalveolar septa. This relationship was unknown. To conclude this chapter, I shall review some recent new findings on this subject.

First, consider the number of arterioles and venules relative to the number of alveoli. Horsfield (1978) has assumed that each alveolus is supplied by one arteriole and drained by one venule. By histological measurements, however, Zhuang et al. (1985) found that each precapillary vessel with a diameter less than 100 μm (arteriole) supplies, on the average, 24.5 pulmonary alveoli, and each venule with diameter less than 100 μm drains, on the average, 17.8 alveoli. This is to say that the total number of alveoli in a given volume divided by the number of arterioles in that volume is 24.5, and the total number of alveoli divided by the number of venules is 17.8. We know, however, that the groups of alveoli supplied by the arterioles are located in separate regions of space than those drained by venules. In fact, Sobin et al. (1980) have shown that in a plane cross-section of the lung parenchyma, the areas occupied by arterioles appear as "islands," whereas the areas occupied by the venules appear as an "ocean" surrounding the isolands. In the cat lung inflated to a transpulmonary pressure of 10 cm H_2O, the characteristic "diameter" of the arteriolar zone is 0.918 ± 0.156 (SD) mm; the characteristic "width" of the venous zone is 1.158 ± 0.410 mm (Sobin et al., 1980). The diameter of the "equivalent sphere" of the alveolus is 163 μm (Zhuang et al., 1985). Thus the diameter of the order 1 polyhedra, which is about three times the alveolar diameter, is ~ 490 μm, and the diameter and width of the arterial and venous zones are approximately twice the dimension of the order 2 polyhedra.

Since the volume of the arterial zone is $43 \pm 7.5\%$ of the total volume of the lung parenchyma (Sobin et al., 1980), the number of alveoli *directly* supplied by the arterioles is 43% of the total, and each arteriole supplies only $0.43 \times 24.5 = 10.5$ alveoli. Similarly, the number of alveoli *directly* drained by each venule is $(1 - 0.43) \times 17.8 = 10.1$ alveoli for the cat.

Further, since the basic unit of respiration is the order 2 polyhedron, each of which has 14 ventilated alveoli, we see that each arteriole *directly* supplies 0.75 order 2 polyhedrons, whereas each venule *directly* drains 0.72 order 2 polyhedrons.

Another finding given in Sobin et al. (1980) is the average path length, L, of blood flowing out of an arteriole and into a venule. It was found to be 556 ± 286 μm in the cat, or approximately one diameter of the order 2 polyhedron. This is the L involved in Eq. 13.6.

Next, consider the structure of the alveolar duct. This much photographed entity has been singularly refractive to three-dimensional modeling. Miller (1947), Orsos (1936), Von Hayek (1960), Weibel (1963), and Wright (1961) have presented morphological observations. Hansen and Ampaya (1975) and Hansen et al. (1975) have reconstructed a physical model of the human terminal bronchiole from histological slides. Dale et al. (1980), Hoppin and Hildebrandt (1977), Karakaplan et al. (1980), Stamenovic and Wilson (1985), and Wilson and Bachofen (1982) have proposed mathematical models, but have not shown that these models agree in quantitative details with the morophometric data published by earlier authors.

In view of this situation, Fung (1988) proposed to begin with the assumptions that (1) all alveoli are equal and space filling, (2) they are ventilated to ducts as uniformly as possible, and (3) they are reinforced at the edges of the ventilation holes (mouths) for structural integrity and distorted by lung weight and inflation according to the theory of elasticity. These hypotheses, coupled with the histological observations that the interalveolar septa are not all triangles, nor all squares, nor all pentagons, led, by elimination, to the proposal that the alveoli are tetrakaidecahedrons (14-hedra of the first order) and the basic unit of alveolar duct is a 14-hedron of the second order (see Fig. 13.11). A order 2 14-hedron is composed of fourteen 14-hedron surrounding a central perforated chamber. Several order 2 14-hedra can be assembled together to form ducts by perforating a few more membrane, as illustrated in Figures 13.12 and 13.13. A assemblage of order 2 and order 1 14-hedra is space filling.

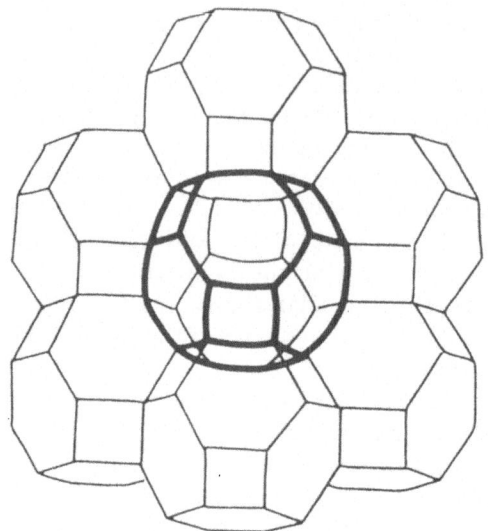

FIGURE 13.11. Sketch of an order 2 polyhedron with fourteen 14-hedra surrounding one central 14-hedron. Several 14-hedra in front are removed to reveal central 14-hedron. To form a basic unit of lung structure, all faces of central 14-hedron are removed, whereas all its edges are reinforced to take up load and keep structure in stable equilibrium. (From Fung, 1988.)

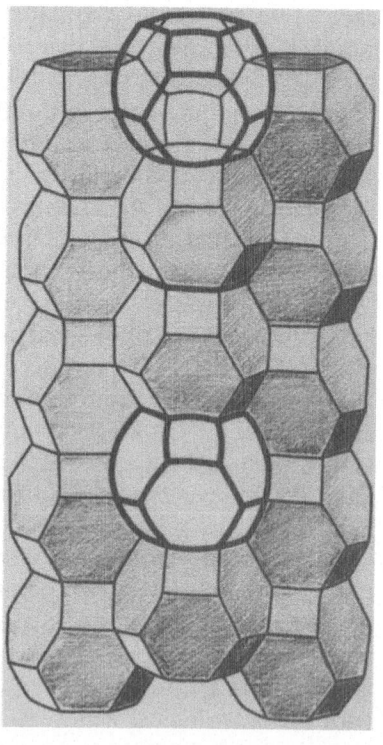

A

B

FIGURE 13.12. *A*: Two order 2 polyhedra joined together to form an alveolar duct. At least one additional face must be removed in order to ventilate the alveoli. This can be identified as the way alveolar ducts of generations 6 and 7 are formed. Ducts of generations 1, 2, and 3 are formed by joining one more order 2 polyhedron to the one shown here (see Figure 13.13). *B*: Another way two order 2 polyhedra can be joined together to form a duct. This can be identified as the way ducts of generations 4 and 5 are formed. (From Fung, 1988.)

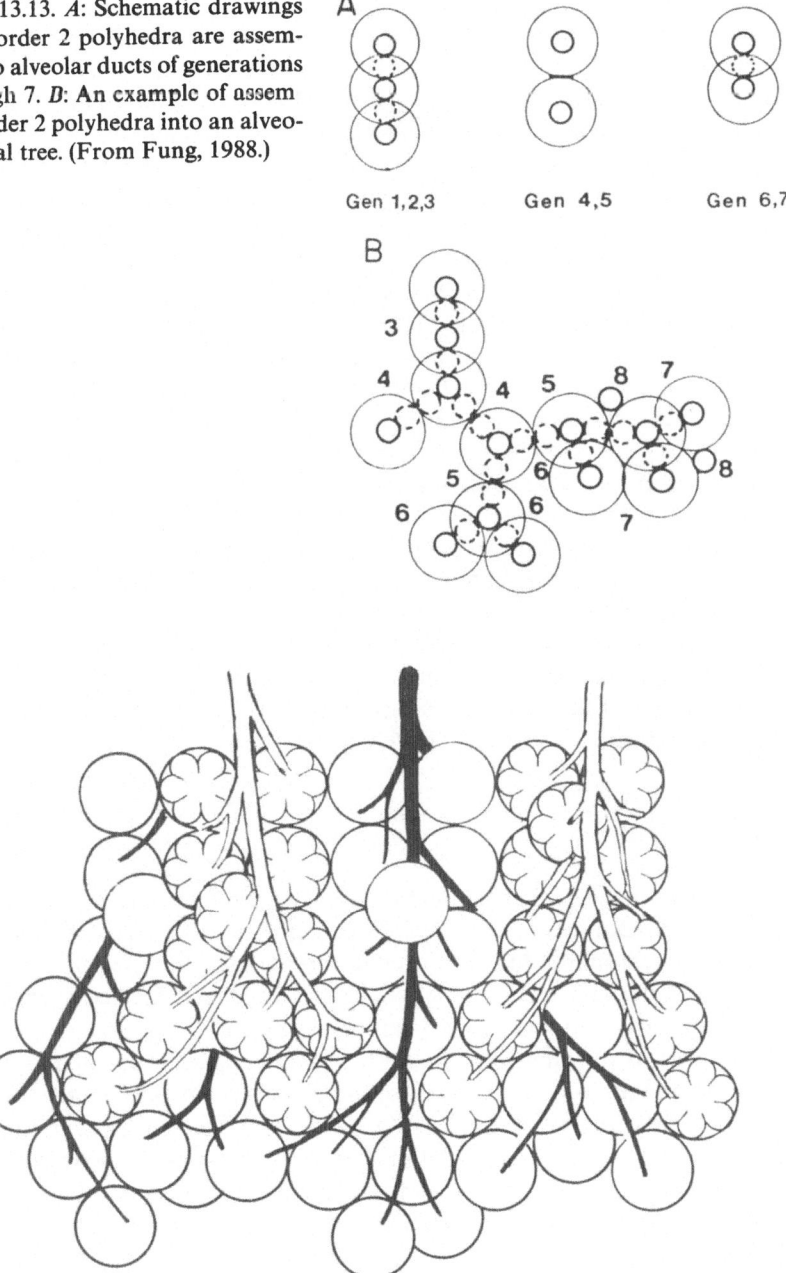

FIGURE 13.13. *A*: Schematic drawings of how order 2 polyhedra are assembled into alveolar ducts of generations 1 through 7. *B*: An example of assembling order 2 polyhedra into an alveolar ductal tree. (From Fung, 1988.)

Gen 1,2,3   Gen 4,5   Gen 6,7

FIGURE 13.14. Conceptual illustration of how circulation and respiration systems are joined at microvascular level. Arterioles are indicated by white color. Venules are indicated by black color. Alveoli are indicated by small loops. Order 2 14 hedra are indicated by larger circles surrounding those loops. Alveoli in venular region are not drawn in. See text for details.

Perforation of a membrane calls for reinforcement of the edges for structural integrity. With elastic deformation taken into account, this model fits the morphometric data of Hansen and Ampaya (1975) and Hansen et al. (1975) quite well, as well as new data by Fung (1988) and Oldmixon et al. (1988) concerning alveolar shapes, alveolar mouth geometry, lengths of alveolar sacs and ducts of successive generations, dihedral angles between interalveolar septa, patterns of alveolar walls in plane cross-sections of lung parenchyma, the types of the vertices, and the ratio of the length of alveolar mouths to the surface area of the alveolar wall. The validation is presented in Fung (1988).

The picture that emerges from this array of quantitative anatomical studies is illustrated schematically in Figure 13.14. The alveoli are represented by small loops inside circles in the arterial regions, and empty circles in the venous regions. The basic units of ducts, the order 2 14-hedra are indicated by larger circles. Alveolar ducts of generations 1 to 7 are composed of order 2 polyhedra, and ducts of generation 8 connect isolated alveolis to longer ducts. Every order 2 polyhedron in an arterial region is perfused by 1.3 terminal arterioles (arteries of order 1); every order 2 polyhedron in the venous region is drained by 1.4 terminal venules (veins of order 1). The widths of the arterial and venous regions are both approximately twice the diameter of the order 2 polyhedron. The arterial regions are male shaped; the venous region is female shaped, having one continuous phase that envelops all arterial regions.

Figure 13.14 illustrates schematically how the tops of the three trees of airway, arteries, and veins of the lung are tied together. It completes the picture shown in Figure 13.1 with which we began.

Conclusions

In this chapter the similarities and differences of pulmonary micromechanics and macromechanics are discussed. Together they make it possible to identify a mathematical model that can predict the pressure-flow relationship of the lung in considerable detail. These predictions are then subjected to experimental validation. A validated model can be used for applications.

The necessity of connecting the studies of microcirculation and macrocirculation in the lung is obvious because these two circuits are in fact connected in series. Nevertheless, the necessity of treating the microcirculation in as much detail as the macrocirculation had not been obvious. For example, in windkessel theory the microcirculation is replaced by a single resistor. In some other theories the microcirculation is represented by a resistance-capacitance circuit. The contention of ths chapter is to show that the microcirculation is as rich in details as the macrocirculation.

Joining the microcirculation to the macrocirculation calls attention to the intermediate regime in which the mathematical simplicity of low Reynolds number in microcirculation (by which the transient and convective inertial force can be neglected) and high Reynolds number in macrocirculation (by which the viscous force can be neglected) is lost. I believe that a thorough study of this regime will be important to the understanding of ventilation-perfusion inequality, nervous control, and drug effects. In this chapter, I attempted only the first step: to clarify the geometric relationship between the ducts, arterioles, and venules.

Limitation of space would not allow me to discuss the applications of the mathematical model presented above; but it would be useful to point out the relationship of this chapter to other chapters. In chapter 12, Dr. Yen presents data on the elasticity of pulmonary arterioles and venules of man. This will enlarge our much needed data base. Currently a complete set of geometric and elasticity data exists only for the cat. For the human lung, geometric data have been presented by Cumming, Horsfield, Weibel, and others. If a complete set of elasticity data were obtained, then a detailed model could be constructed. In Chapter 10, Drs. J.S. and L.P. Lee present an exquisite method to study the mechanics of pulmonary capillaries. In Chapter 8 and 9, Drs. Dawson, Linehan, Bronikowski, and Rickaby present two methods to determine the sites of vasoactivity in pulmonary circulation, one by rapid occlusion and another by a low-viscosity bolus. In Chapter 7, Dr Barman et al. present a new method of measuring pulmonary capillary pressure in vivo. To these new, original contributions, the present chapter may either serve as an introduction, or as an alternative model for comparison. Finally, in Chapter 6, Drs. Schmid-Schönbein, Skalak, and Sutton present a study of blood flow in skeletal muscle in a spirit coinciding with this one. We look forward to see more organs studied this way.

References

Bergel DH, Milnor WR (1965) Pulmonary vascular impedance in the dog. *Circ Res* 16:401–415.

Caro CG, Harrison GK, Mognoni P (1967) Pressure wave transmission in the human pulmonary circulation. *Clin Sci* 23:317–329.

Dale PJ, Mathews FL, Schroter R (1980) Finite element analysis of lung alveolus. *J Biomech* 13:856–873.

Fung YC (1984) *Biodynamics: Circulation.* Springer-Verlag, New York.

Fung YC (1988) A model of the lung structure and its validation. *J Appl Physiol* 64(5):2132–2141.

Fung YC, Sobin SS (1972a) Elasticity of the pulmonary alveolar sheet. *Circ Res* 30:451–469.

Fung YC, Sobin SS (1972b) Pulmonary alveolar blood flow. *Circ Res* 30:470–490.

Fung YC, Sobin SS, Tremer H, Yen MRT, Ho HH (1983) Patency and compliance of pulmonary veins when airway pressure exceeds blood pressure. *J Appl Physiol* 54:1538–1549.

Fung YC, Yen RT (1986) A new theory of pulmonary blood flow in zone 2 condition. *J Appl Physiol* 60:1638–1650.

Fung YC, Zhuang FY (1986) An analysis of the sluicing gate in pulmonary blood flow. *J Biomech Eng* 108:175–182.

Hansen JE, Ampaya EP (1975) Human sir space shapes, sizes, areas, and volumes. *J Appl Physiol* 38:990–995.

Hansen JE, Ampaya EP, Bryant GH, Navin JJ (1975) The branching pattern of airways and air spaces of a single human terminal bronchiole. *J Appl Physiol* 38:983–989.

Hoppin FG Jr, Hildebrandt J (1977) Mechanical properties of the lung. In West JB (ed) *Bioengineering Aspects of the Lung.* (Lung Biology in Health and Disease Series, Vol 3.) Marcel Dekker, New York, pp 83–162.

Horsfield K (1978) Morphometry of the small pulmonary arteries in man. *Circ Res* 42:593–597.

Horsfield K, Gordon WI (1981) Morphometry of pulmonary veins in man. *Lung* 159:211–218.

Karakaplan AD, Bieniek MP, Skalak R (1980) A mathematical model of lung parenchyma. *J Biomech Eng* 102:124–136.

Krahl VW (1964). Anatomy of mammalian lung. In Fenn WO, Rahn H (eds) *Handbook of Physiology*, Sec 3, *Respiration*, Vol 1. American Physiological Society, Washington, DC, pp 213–284.

Lee JS (1969) Slow viscous flow in a lung alveoli model. *J Biomech* 2:187–198.

Lee JS, Fung YC (1968) Experiments on blood flow in lung alveoli models. ASME Paper No. 68-WA/BHF-2, American Society of Mechanical Engineers, pp 1–8. Presented in Winter Annual Meeting.

Miller WS (1947) *The Lung.* Charles C Thomas, Springfield, IL.

Milnor WR (1972) Pulmonary hemodynamics. In Bergel DH (ed) *Cardiovascular Fluid Dynamics* Vol 2. Academic Press, New York, pp 299–340.

Oldmixon EH, Butler JP, Hoppin FG Jr (1988) Dihedral angles between alveolar septa. *J Appl Physiol* 64:299–307.

Orsos F (1936) Die Grüstsystem der Lunge und deren physiologische und pathologische Bedeutung. *Beitr Klin Tuberk Spezif Tubert Forsch* 87:568–609.

Patel DJ, de Freitas FM, Fry DL (1963) Hydraulic input impedance to aorta and pulmonary artery in dogs. *J Appl Physiol* 18:134–140.

Pedley TJ, Schroter RC, Sudlow MF (1977) Gas flow and mixing in the airways. In: West J (ed) *Bioengineering Aspects of the Lung.* Marcel Dekker, New York, pp 163–265.

Permutt S, Bromberger-Barnea B, Bane HN (1962) Alveolar pressure, pulmonary venous pressure, and the vascular waterfall. *Med Thorac* 19:239–260.

Permutt S, Caldini P, Maseri A, Palmer WH, Sasamori T, Zierler K (1969) Recruitment versus distensibility in the pulmonary vascular bed. In Fishman AP, Hecht HH (eds) *The Pulmonary Circulation and Interstitial Space.* University of Chicago Press, Chicago, pp 375–387.

Permutt S, Riley RL (1963) Hemodynamics of collapsible vessels with tone: The vascular waterfall. *J Appl Physiol* 18:924–932.

Pollack GH, Reddy RV, Noordergraaf A (1968) Input impedance, wave travel, and reflections in the human pulmonary arterial tree: Studies using an electrical analog. *IEEE Trans Biomedical Eng* BME-15:151–164.

Sobin SS, Dung YC, Tremer HM, Rosenquist TH (1972) Elasticity of the pulmonary alveolar microvascular sheet in the cat. *Circ Res* 30:440–450.

Sobin SS, Fung YC, Lindal RG, Tremer HM, Clark L (1980) Topology of pulmonary arterioles, capillaries, and venules in the cat. *Microvasc Res* 19:217–233.

Sobin SS, Lindal RG, Fung YC, Tremer HM (1978) Elasticity of the smallest non-capillary blood vessels in the cat. *Microvasc Res* 15:57–68.

Sobin SS, Tremer HM, Hardy JD, Chiodi HP (1983). Changes in arteriole in acute and chronic hypoxic pulmonary hypertension and recovery in rat. *J Appl Physiol* 55:1445–1455.

Stamenovic D, Wilson TA (1985) A strain energy function for lung parenchyma. *J Biomech Eng* 107:81–86.

von Hayek H (1960) *The Human Lung*. Hafner, New York.

Warrell DA, Evans JW, Clarke RO, Kingaby GP, West JB (1972) Pattern of filling in the pulmonary capillary bed. *J Appl Physiol* 32:346–356.

Weibel ER (1963). *Morphometry of the Human Lung*. Academic Press, New York.

Wiener F, Morkin E, Skalak R, Fishman AP (1966) Wave propagation in the pulmonary circulation. *Circ Res* 19:834–850.

Wilson TA, Bachofen H (1982) A model for mechanical structure of the alveolar duct. *J Appl Physiol* 52:1064–1070.

Wright RR (1961) Elastic tissue of normal and emphysematous lungs. A tri-dimensional histologic study. *Am J Pathol* 39:355–366.

Yen MRT, Foppiano L (1981) Elasticity of small pulmonary veins in the cat. *J Biomech Eng Trans ASME* 103:38–42.

Yen MRT, Fung YC (1973) Model experiments on apparent blood viscosity and hematocrit in pulmonary alveoli. *J Appl Physiol* 35:510–517.

Yen MRT, Fung YC, Bingham N (1980) Elasticity of small pulmonary arteries in the cat. *J Biomech Eng* 102:170–177.

Yen RT, Fung YC, Zhuang FY, Zeng YJ (1984a) Comparison of theory and experiments of blood flow in cat's lung. In Fung YC, Fukada E, Wang JJ (eds) *Biomechanics in China, Japan, and U.S.A.* Science Press, Beijing, China, pp 240–253.

Yen RT, Sobin SS (1986) Pulmonary blood flow in the cat: Correlation between theory and experiment. In Schmid-Schönbein GW, Woo SL-Y, Zweifach BW (eds) *Frontiers in Biomechanics*. Springer-Verlag, New York, pp 365–376.

Yen MRT, Zhuang FY, Fung YC, Ho HH, Tremer H, Sobin SS (1983) Morphometry of the cat's pulmonary venous tree. *J Appl Physiol* 55:236–242.

Yen MRT, Zhuang FY, Fung YC, Ho HH, Tremer H, Sobin SS (1984b). Morphometry of the cat's pulmonary arterial tree. *J Biomech Eng* 106:131–136.

Zhuang FY, Fung YC, Yen MRT (1983) Analysis of blood flow in cat's lung with detailed anatomical and elasticity data. *J Appl Physiol* 55:1341–1348.

Zhuang FY, Yen MRT, Fung YC, Sobin SS (1985) How many pulmonary alveoli are supplies by a single arteriole and drained by a single venule? *Microvasc Res* 29:18–31.

Index